SUBCELLULAR PARTICLES, STRUCTURES, AND ORGANELLES

METHODS IN MOLECULAR BIOLOGY

Edited by

ALLEN I. LASKIN
ESSO Research and Engineering Company
Linden, New Jersey

JEROLD A. LAST
Harvard University
Cambridge, Massachusetts

VOLUME 1: Protein Biosynthesis in Bacterial Systems,
edited by Jerold A. Last and Allen I. Laskin

VOLUME 2: Protein Biosynthesis in Nonbacterial Systems,
edited by Jerold A. Last and Allen I. Laskin

VOLUME 3: Methods in Cyclic Nucleotide Research,
edited by Mark Chasin

VOLUME 4: Nucleic Acid Biosynthesis,
edited by Allen I. Laskin and Jerold A. Last

VOLUME 5: Subcellular Particles, Structures, and Organelles,
edited by Allen I. Laskin and Jerold A. Last

VOLUME 6: Immune Response at the Cellular Level,
edited by Theodore P. Zacharia

VOLUME 7: DNA Replication and Biosynthesis
edited by Reed Wickner
(in preparation)

SUBCELLULAR PARTICLES, STRUCTURES, AND ORGANELLES

EDITED BY

Allen I. Laskin and Jerold A. Last

ESSO Research and Engineering Company
Linden, New Jersey

Harvard University
Cambridge, Massachusetts

MARCEL DEKKER, INC. New York 1974

MARCEL DEKKER, INC.
305 East 45th Street, New York, New York 10017

LIBRARY OF CONGRESS CATALOG CARD NUMBER: 73-90306
ISBN: 0-8247-6067-0

Current printing (last digit):
10 9 8 7 6 5 4 3 2 1

PRINTED IN THE UNITED STATES OF AMERICA

PREFACE

In this volume of the "Methods in Molecular Biology" series an attempt is made to provide some representative methodology for use in studying various subcellular structures and organelles. The methods presented here, of course, represent only a small sampling of the vast number that could have been chosen. Hopefully, however, there is enough variety and range to provide useful information for working with other systems that are not specifically detailed here.

Once again, as in the previous volumes in the series, the authors were asked to prepare critically written descriptions of methods used in a specific area, and whenever appropriate, to discuss such things as: why a particular approach was taken; why a particular reagent was used; what alternatives are feasible and acceptable; what to do "if things go wrong"; etc.

The book begins with a chapter on bacterial ribosomes, which details their isolation and the extracting, purification, analysis, and identification of ribosomal proteins. Reconstitution of 30 S ribosomes from their RNA and protein components is also described. The second chapter covers another bacterial structure--flagella, and details the purification of intact flagella as well as flagellar components (flagellar filaments, flagellin, flagellar hooks, and hook-basal body complexes).

Chapter 3 begins the section on mammalian cells, and presents isolation procedures for plasma membranes, especially as applied to cell surface receptor work. Next follow chapters on the isolation of lysosomes and of Golgi apparatus. The final three chapters cover more general methodology. First, there is a detailed description of zonal centrifugation, which has found wide applicability in the isolation of subcellular materials. Then, two separate chapters cover methods in electron microscopy that are broadly applicable.

As can be expected, there is a degree of overlap in these presentations, as well as with material in other volumes in this series. We feel that this is not only an inevitable, but a useful consequence of the way in which these methods are presented, since it provides the reader with alternative viewpoints and techniques for his consideration.

ALLEN I. LASKIN
JEROLD A. LAST

CONTRIBUTORS TO THIS VOLUME

SYDNEY S. BREESE, JR., Plum Island Animal Disease Laboratory, Veterinary Sciences Research Division, Agricultural Research Service, United States Department of Agriculture, Greenport, New York; and National Academy of Sciences, 2101 Constitution Avenue, Washington, D.C.

GARY R. CRAVEN, Laboratory of Molecular Biology, University of Wisconsin, Madison, Wisconsin

WILLIAM P. CUNNINGHAM, Department of Genetics and Cell Biology, University of Minnesota, St. Paul, Minnesota

M. L. DePAMPHILIS*, Department of Biochemistry, University of Wisconsin, Madison, Wisconsin

C. RONALD KAHN, Section on Biophysical Chemistry, Laboratory of Neurochemistry, National Institute of Mental Health, National Institutes of Health, Bethesda, Maryland; and Diabetes Section, Clinical Endocrinology Branch, National Institute of Arthritis, Metabolism and Digestive Diseases, National Institutes of Health, Bethesda, Maryland

*Present address: Department of Biological Chemistry, Harvard Medical School, Boston, Massachusetts

DAVID M. NEVILLE, JR., Section on Biophysical Chemistry, Laboratory of Neurochemistry, National Institute of Mental Health, National Institutes of Health, Bethesda, Maryland; and Diabetes Section Clinical Endocrinology Branch, National Institute of Arthritis, Metabolism and Digestive Diseases, National Institutes of Health, Bethesda, Maryland

C. A. PRICE, Department of Biochemistry and Microbiology, Rutgers University, New Brunswick, New Jersey

GERALD L. ROWIN, Wallace Laboratories, Cranbury, New Jersey

THEODORE P. ZACHARIA, Plum Island Animal Disease Laboratory, Veterinary Sciences Research Division, Agricultural Research Service, United States Department of Agriculture, Greenport, New York; and National Academy of Sciences, Washington, D.C.

CONTENTS

SUBCELLULAR PARTICLES, STRUCTURES, AND ORGANELLES

Chapter 1

THE STRUCTURE OF THE BACTERIAL RIBOSOME

Gary R. Craven

Laboratory of Molecular Biology
University of Wisconsin
Madison, Wisconsin

I. INTRODUCTION

The macromolecular components involved in protein biosynthesis and their sequence of action can now be confidently listed. This amounts to a biochemical pathway of protein formation. However, almost nothing concrete is known about the molecular details of the various steps in the pathway.

The most glaring weakness in our understanding of the mechanism of protein synthesis centers around the structure and function of the ribosome. Ribosomes are complicated assemblages of about fifty chemically distinct proteins [1-3] and three molecules of RNA. Beyond this almost nothing is known. Little specific information is available dealing with the three-dimensional arrangement of all these macromolecules, and only a few experiments have been designed to identify the role of the various ribosomal components in protein synthesis.

However, we are on the threshold of developing a complete three-dimensional picture of the ribosome correlated with its many functions. (1) In the last few years, techniques have been developed that should promote the investigation of the structure and function of the ribosomes in molecular detail. For example, all the proteins composing the 50S and the 30S ribosomes have been purified and characterized. (2) Relatively simple techniques of analysis are available that make it possible to identify each protein and to correlate such identifications with those of other workers in the field [4]. (3) Finally, it is now possible to reassemble *in vitro* the ribosomes and all intermediate precursor particles from purified RNA and the individual proteins [5]. These three techniques open many new approaches to the problems of ribosome structure and function. Thus, this chapter will not deal specifically with the structure of the ribosome, but in fact will attempt to describe the experimental details involved in these three methods of analysis.

II. PREPARATION OF RIBOSOMES

For studies involving ribosome structure, either *Escherichia coli*, strain MRE 600, or *E. coli* K_{12}, strain Q_{13}, is definitely recommended. Both of these strains have very low levels of RNase activity (Q_{13} is also deficient in polynucleotide phosphorylase). The MRE 600 has the additional advantage that it grows exceedingly well in minimal medium. However, for studies involving genetic analysis, the Q_{13} strain is preferred.

There are principally two methods presently in use for ribosome preparation. The original method of Tissieres et al. [6], involving differential centrifugation, is applied primarily in investigations concerned with the function of the ribosome. This method yields a ribosome fraction that is highly active in all phases of protein synthesis, but that is complicated by the presence of a multitude of unwanted protein contaminants. These contaminating supernatant proteins can be preferentially removed by washing the ribosomes with buffers containing high concentrations of salt.

Two procedures have been developed for washing with salt; one uses ammonium chloride and the other ammonium sulfate. The ammonium chloride procedure was developed explicitly to release the initiation factors, whereas the ammonium sulfate procedure uses a lower concentration of salt and was developed to prepare a ribosome fraction containing a reproducible protein component [7]. Most importantly, it has been demonstrated that this protein component is itself not affected by the washing with salt. Thus, this procedure is definitely preferred until methods are developed to better define ribosomal proteins. For these reasons, this procedure is described here.

To start, 100 g of *E. coli* cell paste is suspended in 120 ml of TM buffer [10 mM Tris·HCl (pH 8.0), 10 mM Mg $(acetate)_2$] along with 8 ml of 10-times-concentrated TM buffer and 500 μg of DNase. The cell suspension is disrupted by alumina grinding or by passage through a French press, and diluted to 275 ml with TM buffer. The crude extract is centrifuged at 30,000*g* for 30 min. The supernatant is precipitated by the slow addition of 57.7 g of ammonium sulfate. The precipitate is removed by centrifugation, and the ribosome fraction is subsequently precipitated by the addition of 57.7 g of ammonium sulfate to the supernatant. The resultant precipitate should contain most of the ribosome fraction. It is then dissolved in 275 ml of TM buffer, and the ammonium sulfate fractionation procedure is repeated. The addition of the first batch of ammonium sulfate should be carried out with caution. Residual ammonium sulfate is present in the precipitate and, thus, some ribosomes may precipitate during the first addition of salt. This unwanted precipitation can be avoided by monitoring the absorbance of the supernatant at 260 nm. Usually less than 55 g of ammonium sulfate is required in this first precipitation.

The final ribosome precipitate from the second salt fractionation is dissolved in 275 ml of TM buffer, and the salt fractionation procedure is repeated a third time. The final precipitate is dissolved in 350 ml of TM buffer containing 78 g/liter (0.6 M) of ammonium sulfate. The ribosomes are then pelleted by centrifugation at 30,000 rpm for 10 hours. After centrifugation the tubes are placed upside down in the cold to allow a loose layer of material of unknown composition to slide down the side of the tube. This material is removed with a rubber policeman and the pellets are suspended in 350 ml of TM buffer containing 78 g/liter of ammonium sulfate. Any undissolved material is removed by low-speed centrifugation, and the ribosome fraction is again pelleted by centrifugation at 30,000 rpm for 10 hours. The ribosomal pellet, which should be clear and colorless, is suspended in a minimal volume of TM buffer, dialyzed against the same buffer, and stored at -70°C. It is extremely important to dialyze the ribosomal preparation at this step as it contains a significant concentration of residual NH_4^+ ions.

The ribosomes prepared in this manner are highly active. They exist as a mixture of free 50S and 30S subunits. The yield should be around 35,000 A_{260} units (2.3 g) from 100 g of cell paste. The 50S and 30S subunits are separated by differential centrifugation or by zonal centrifugation after dialysis against TKM buffer [10 mM Tris·HCl (pH 8.0), 50 mM KCl, 0.5 mM Mg $(acetate)_2$].

Many structural studies require the absolute absence of contaminating nuclease activity. Although the strains of *E. coli* suggested here for use are RNase I free, significant nuclease activity may be observed in the final preparation. This activity is apparently owing to contamination from skin, chemicals, and glassware used during the purification procedure. Such contamination can be avoided by the use of gloves at all times and by treatment of glassware with diethylpyrocarbonate. The diethylpyrocarbonate is a general cross-linking reagent that readily inactivates nuclease activities. The glassware is soaked in a 1% solution of the reagent for 30 min. It is then rinsed and autoclaved. Diethylpyrocarbonate decomposes under the influence of either heat or low pH. In addition, solutions not containing compounds having free amino groups (e.g., sucrose) can be rendered nuclease free by this treatment.

III. EXTRACTION OF PROTEIN

There are currently three alternative methods used to extract the ribosomal proteins. Two of these involve selective precipitation of the RNA component, and the third uses the degradative action of ribonuclease. The RNA precipitation methods have the advantage that not only is the total protein component obtained in soluble form, but the RNA is preserved as well.

The use of ribonuclease in the liberation of ribosomal proteins has some special interest. Originally Spahr [8] introduced this technique as a convenient means of releasing the latent ribonuclease activity associated with the 30S ribosome. The ribosomes are dissolved in 8 M urea and dialyzed against a convenient buffer. During the dialysis the RNase activity is expressed, digesting the RNA and releasing the protein as an insoluble precipitate. Recently this approach has been modified by the use of RNase-free strains. The ribosomes from these strains do not undergo very rapid autodigestion. However, small amounts of pancreatic RNase readily can be added to digest the RNA. This technique is especially convenient for the polyacrylamide gel analysis of ribosomal proteins, as one can apply whole ribosomes directly to the sample gel, along with urea and RNase. Thus, one obtains a complete gel pattern of the ribosomal proteins without using an indirect extraction procedure.

Recently, an adaptation of the RNase method has been reported [9] that is designed to extract the ribosomal proteins without the use of severe conditions of denaturation such as urea or of high concentrations of acetic acid. Dialysis of ribosomes against 1 M NH_4Cl and low Mg^{++} unfolds the ribosomes and allows the RNase activities associated with the preparation to be expressed [9]. (We have found that unpurified MRE 600 ribosomes work well in this system, indicating that RNase activities other than RNase I are responsible.) As the RNA is digested the protein precipitates; the protein can be redissolved to yield an extract essentially unexposed to harsh denaturants. This method has not been systematically investigated and will therefore not be detailed here. It does have potential merit, however, and we hope that it will be widely explored.

The method of protein extraction we find most useful in our laboratory uses 66% acetic acid, as originally developed by Waller and Harris [10]. The method as applied to ribosomes was studied by Hardy et al. [11], and was found to extract more than 95% of the protein. The method is convenient and simple and allows the rapid extraction of protein under conditions that preclude the formation of many types of artifacts. Another advantage of this procedure is that in certain instances it can be used as a means of concentration of the extracted protein. Thus, the ribosome sample can be extracted by dialysis against a large volume of 66% acetic acid containing 0.03 M $MgCl_2$. With this technique the RNA precipitates inside the dialysis bag, the protein is solubilized, and the volume of the sample is reduced by a factor of about three. If desired, the volume of the sample can be reduced even further by lyophilization. We have found this approach very useful in the analysis of fractions from a sucrose gradient, where it is necessary to remove the sucrose, extract the protein, and concentrate the sample all in one step.

If direct extraction of the protein is desired, the procedure recently outlined by Hardy et al. [11] is recommended. In this method, 2 vol of ice-cold glacial acetic acid are slowly added with vigorous agitation to the ribosome suspension, previously made 0.1 M in Mg^{2+}. The solution is periodically agitated; after 30 min it is centrifuged. The RNA pellet is resuspended in 66% acetic acid and again centrifuged. The two acetic acid supernatants are combined and dialyzed against the appropriate buffer.

The third method currently followed uses high concentrations of LiCl to precipitate the RNA in the presence of 8 M urea, which keeps the proteins soluble. This method has been widely used, but does not offer any significant advantage over the acetic acid procedure.

IV. ANALYSIS AND IDENTIFICATION OF THE PROTEINS

If the procedures outlined in Section III for the preparation of ribosomes and the extraction of the protein are followed, about 21 chemically distinct proteins can be identified as components of the 30S subunit, and about 29 are found to be associated with the 50S particle. The problem faced by all laboratories dealing with the structure of the ribosome is the identification of these some 50 components. A recent development in the technology of gel electrophoresis has made it possible to identify almost all the ribosomal proteins (both 50S and 30S) and to correlate the identifications with those of other laboratories [12].

Many of the investigations of ribosome structure have used a single polyacrylamide gel electrophoresis for the analysis of results. This system is that originally designed by Leboy et al. [13]. Unfortunately, with this procedure only 7 of the 21 type 30S proteins and only 13 of the 29 type 50S proteins are resolved into unique components. Thus, any analysis of ribosome structure or function that uses this procedure exclusively inevitably results in considerable ambiguity. A modification introduced by Voynow and Kurland [14] improves the situation a bit; it allows the resolution of 5 additional 30S components. This modification is a valuable addition to the ribosome chemist's arsenal, and is highly recommended. However, a method is needed that achieves complete separation of virtually all ribosomal protein components. Polyacrylamide electrophoresis in two dimensions, under the conditions prescribed by Kaltschmidt and Wittmann [12], meets this requirement.

We have adapted the two-dimensional fingerprinting technique in our laboratory and have found it to be extremely useful and highly reproducible. Although our method of preparation of ribosomes is different from that originally described [12], we have been able to readily identify all the 30S proteins and most of the 50S proteins that we have studied.

Although two-dimensional electrophoresis of the ribosomal proteins is an important advance, it does have two outstanding disadvantages that should be discussed. The first is more of an inconvenience than a disadvantage. By necessity, the apparatus for running the second dimension can accommodate a limited number of samples (five in the design of Kaltschmidt and Wittmann). A complete run of both dimensions and staining and destaining of the gels take about four days from start to finish. These factors seriously limit the number of samples that can be conveniently analyzed without a major investment in personnel and equipment.

The second disadvantage is of a more serious nature. In order to detect all components, a relatively large quantity of protein is required. For the complete subunits, between 50 and 100 $\underline{A}_{260}$ units must be applied to a single gel. (This is roughly

20 times the amount required in the analytical system.) For individual proteins, between 50 and 100 μg is needed for each analysis. For many experiments this requirement is a definite limitation. However, until a better approach is developed, two-dimensional gel electrophoresis should be used for the correlation and identification of the ribosomal proteins. This technique will, therefore, be outlined here.

In the first dimension, the sample is placed in the middle of a gel of running pH 8.6. This pH is close to the apparent isoelectric point of most of the basic proteins, and affords a separation predominantly on the basis of charge. The gel components are given in Table 1.

TABLE 1

Composition of First-Dimension Gels

	Grams in		
Chemical	Running gel	Sample gel	Electrode buffer
Urea	54.0	48.0	360.0
$EDTA\text{-}Na_2$	1.2	0.085	2.4
Boric acid	4.8	0.32	9.6
Tris	7.3	--	14.55
N,N,N,N-tetra-methylene diamine (TEMED)	0.45 ml	0.06 ml	--
Acrylamide	6.0	4.0	--
Bisacrylamide	0.2	0.2	--
Water, make up to	148.5 ml	99.0 ml	1000 ml

A section of glass tubing 180 by 6 mm is sealed at one end with the inside portion of a serum stopper. The running gel is mixed with ammonium persulfate (final concentration 0.07%) and added to the tube to a level of 70 mm. The gel solution is overlayered with water; after polymerization is completed the water is removed with absorbent paper. This process creates a smooth surface on the running gel.

The sample is dissolved in sample gel. Kaltschmidt and Wittmann [12] suggest that the sample should be lyophilized and dissolved directly in the sample gel solution. It is sometimes more convenient to mix the sample with RNase and urea (final concentration 8 M) and then to mix with sample gel, thus avoiding the lyophilization step. Good polymerization is achieved if the volume of the sample is no greater than 50 μl. The final volume should not exceed 150 μl. Thus, the procedure outlined in Section III for the extraction of protein (i.e., dialysis against 66% acetic acid followed by lyophilization) is well suited for this application. The sample is polymerized by the addition of riboflavin (final concentration, 5 μg/ml) and ammonium peroxodisulfate (final concentration, 50 μg/ml) and by exposure to intense light. The sample gel is also overlayered with water before polymerization, and the water is subsequently removed with absorbent paper.

After polymerization of the sample gel, the tube is filled with running gel. Thus, the sample is embedded in the middle of two sections of running gel. Electrophoresis of this gel is accomplished with a voltage of 90 V, at an initial amperage of 2.5 mA per tube. Constant current is maintained throughout the run. The run is conducted for 24 hours, with the electrode buffers refreshed after 15 hours to remove accumulated persulfate and N,N,N,N-tetramethylethylenediamine. The gels are removed after the run and soaked in the buffer used in the second dimension, as detailed in Table 2.

TABLE 2

Composition of Second-Dimension Gels

Chemical	Grams in Running gel	Soaking gel	Electrode buffer
Urea	360.0	480.0	--
Glacial acetic acid	52.3 ml	0.74 ml	15.0 ml
KOH	9.6 ml of 5 N	2.4 ml of 5 N	--
Acrylamide	180.0	--	--
Bisacrylamide	5.0	--	--
TEMED	5.8 ml	--	--
Water, make up to	967.0 ml	1000 ml	1.0 liter

The gels are equilibrated with the second-dimension buffer by soaking in the starting buffer for three hours, with agitation. The authors of this technique refer to this process as dialysis. The gel is then layered over a slab of the second-dimension gel (running pH 4.6, 18% acrylamide). The gel is embedded by the addition of running gel. The dimensions of the second-dimension slab are 200 x 200 x 10 mm. The construction of the apparatus used for the second dimension will not be discussed here. We have reproduced the apparatus designed by Kaltschmidt and Wittmann and found it to be generally satisfactory. The voltage applied is a constant 105 V, with a current of 480 mA per five slabs. The running time is about 24 hours, and is determined by the migration of a tracking dye (e.g., methyl green) that is included in the top electrode buffer. The resultant slabs can be stained by either amino black or Coomassie brilliant blue. The latter is somewhat more sensitive but has the disadvantage of requiring a 10% trichloroacetic acid (TCA) solvent, which causes the slab to swell extensively.

We have found that the method is so reproducible that identifications can be made merely by a direct comparison with a fingerprint of a complete protein mixture. However, there may be occasional ambiguity, in which case the total protein can be included with the protein(s) to be identified at 0.1 times the concentration of the unknown.

One note of caution should be expressed concerning the interpretation of two-dimensional fingerprints. Several of the proteins have sulfhydryl groups, which can readily oxidize at the pH of the first-dimension gel. This can lead to the appearance of new spots on the fingerprint. To avoid this artifact, it is good practice to carboxymethylate the protein extract with iodoacetic acid before the first dimension is run. This is best achieved by dissolution of the protein after lyophilization from acetic acid in an 8-M urea solution containing a three-fold excess of iodoacetic acid over the calculated number of available SH groups. The urea is buffered with 0.1 M Tris, at a pH of 8.2. After 20 min of incubation, the reaction is terminated by the addition of a five-fold excess of 2-mercaptoethanol. The pH should be checked during the reaction and adjusted if it falls below 8.0. Carboxymethylation does not appear to alter the migration of any proteins except those that are artifacts produced by oxidation. If the protein concentration is high enough, the sample can be directly applied to the first-dimension gel. The reactants can be removed, if necessary, by dialysis against 50% acetic acid, followed by lyophilization.

V. PURIFICATION OF RIBOSOMAL PROTEINS

Many investigations of the structure and function of ribosomes will inevitably require an ample supply of the individual proteins in pure form. Two slightly different column procedures have been developed that afford nearly complete purification of all the 30S proteins, and separate most of the 50S proteins into preliminary groups. Both procedures use cellulose ion exchangers and are very similar in most aspects. We find the procedure by Hardy et al. [11], which uses phosphocellulose, to be the more suitable for the large-scale purification of the 30S proteins, whereas the procedure published by Moore et al. [14a], which uses carboxymethylcellulose, is particularly well suited for analytical analysis. The method with phosphocellulose, as applied to the separation of 30S proteins, will be described here.

We generally start with between 5000 and 10,000 A_{260} units of 30S ribosomes. The protein is extracted with 66% acetic acid and dialyzed against 6 M urea, 0.05 M phosphate, buffered at pH 5.8. It is extremely important to check the pH of the sample before applying it to the column. The urea buffer also contains 1 mM dithiothreitol (DTT) to prevent the formation of spurious oxidation products.

The column elution profile can be most conveniently monitored by measurement of the absorbance at 230 nm. As most commercial preparations of urea contain impurities that absorb at this wavelength, it is necessary to clarify all of the urea buffer with activated charcoal, as originally described by Hardy et al. [11]. To do this, the urea solution is mixed with activated charcoal (about 1 g per liter of buffer). It is stirred for 4 hours and filtered. If the absorbance at 230 nm is greater than 0.1, then the process should be repeated.

The preparation of the phosphocellulose is similar to that described by Kurland et al. [15]. It is thoroughly washed, alternately, with 0.5 N HCl, 0.5 N NaOH, and 1.0 N NaCl, after which fine particles are removed by decantation from distilled water. It is washed finally with phosphate buffer (pH 5.8), then with the urea-phosphate-DTT column buffer. The washing procedure is best accomplished by removal of the buffer by suction on a Buchner filter. The new buffer is then washed through the phosphocellulose cake until the effluent has the proper pH.

A column with dimensions of 2.5 x 75 cm is packed with the phosphocellulose slurry. An extension of the column should be used, and the slurry should be filled to a height of 150 cm. The column is packed by flow with a gravity head of only 25 cm. The low-gravity head is achieved by running a piece of small-bore tubing from the bottom of the column back up to the top of the extension. A column packed in this manner has excellent flow characteristics and can be used without the aid of a pump or applied pressure.

Once the column is packed to a level of about 65 cm and washed extensively with fresh buffer, the extension can be removed and the protein sample adsorbed. The column is eluted with a 4-liter linear gradient of NaCl, having a limiting concentration of 0.5 M. However, the gradient is not initiated until after the first protein peak has been eluted, which takes about 500 ml of the starting urea buffer. The flow rate is generally about 20 ml per hour. The proteins are identified by a combination of elution position and polyacrylamide gel electrophoresis, as outlined above. Polyacrylamide gel electrophoresis is used also to determine what fractions should be pooled. The pooled fractions can be concentrated with a Diaflow ultrafilter (UM-2) and stored indefintely at -20°C. Further purification of some proteins is usually required, and can be performed by rechromatography or, in some instances, by gel filtration on Sephadex G-100 equilibrated with the same 6 M urea-phosphate-DTT buffer. For details of these procedures, the original description by Hardy et al. [11] should be consulted.

VI. RECONSTITUTION OF RIBOSOMES

The original method developed by Traub and Nomura [5] for the total reconstitution of 30S ribosomes from 16S RNA and extracted ribosomal proteins is now routine. With the availability of bulk quantities of the individual purified proteins as described in the previous section, many questions about the structure and function of ribosomes can be answered with the reconstitution procedure. We have found that the procedure as outlined by Traub et al. [16] is highly reproducible and reliable. As we have incorporated a few slight modifications that make the system somewhat more convenient, our procedure will be described here.

A. Preparation of the 16S RNA

We have found it most convenient to use the RNA precipitated from 30S ribosomes in the acetic acid extraction of the proteins, described in Section II. This RNA, if handled carefully, works with the same efficiency as that prepared by phenol extraction of the 23S core particle, as suggested by M. Nomura.

The RNA precipitated with acetic acid is washed several times with cold buffer (0.01 M tricine, .01 M $MgCl_2$), until the pH of the solution is approximately 7.0. The precipitate is then resuspended in 5 mM phosphate buffer (pH 6.0) and allowed to stand with occasional agitation, at room temperature, until it is completely dissolved. The phosphate buffer and all glassware should be treated with diethylpyrocarbonate to inactivate RNases. To remove any residual protein, we usually extract this RNA solution with phenol, following the procedure of Monier et al. [17]. However, phenol extraction is not essential, as the RNA at this stage is completely active in the reconstitution system.

B. Preparation of the Ribosomal Protein

The supernatant obtained in the acetic acid extraction contains all the ribosomal proteins. This preparation can be used in the reconstitution system with good results, provided that it is first dialyzed against 6 M urea-0.05 M phosphate, pH 6.0. We have tried to dialyze directly into the proper reconstitution buffer without good results. We also find that removal of the acetic acid by lyophilization yields a protein preparation that has greatly reduced reconstitution activity.

C. Procedure for Total Reconstitution

The critically important parameters in the reconstitution process are ionic strength, magnesium concentration, pH, and temperature. The reconstitution buffer consists of 5 mM phosphate (pH 7.5), 0.02 M $MgCl_2$ 0.33 M KCl, and 10 mM β-mercaptoethanol, or 1 mM dithiothreitol. In this buffer, the final ionic strength is 0.37, which is the optimum for the assembly reaction. Incubation in this buffer is usually for 20 min, and the temperature is 42°C. The 16S RNA is stored in the 5 mM phosphate buffer at -70°C, and is adjusted to 0.02 M Mg^{++} before use. It is most convenient to keep the RNA at concentrations greater than 100 A_{260} units per ml, thus allowing the flexibility of directly diluting the RNA into the appropriate buffer. The final concentration of RNA is not critical, but best results seem to be attained if it is kept around 2-5 A_{260} units per ml. The amount of protein used is determined by the RNA concentration, and is at least 1.2 equivalents (an equivalent is defined as that amount of protein extracted from ribosomes equivalent in A_{260} units to the amount of RNA used). In some cases, however, we have used up to 2.5 equivalents of protein to ensure complete efficiency in the formation of complex.

Traub and Nomura [5] found that it is necessary to incubate the RNA at 42°C for 10 min before the addition of the protein. Normally, we dilute the RNA to a final volume of 7 ml with the phosphate buffer containing 0.02 M $MgCl_2$. The protein is dialyzed from urea, as described in the preceding subsection, into a buffer containing 1.0 M KCl, 0.02 M $MgCl_2$, 0.01 M β-mercaptoethanol, and 5 mM phosphate. Then 3 ml of this protein solution is added dropwise, with stirring, to the RNA to yield a final concentration of KCl of 0.33 M.

After incubation, the protein-RNA mixture is centrifuged at 17,000 rpm for 10 min to remove insoluble excess protein. It is then layered over 5 ml of 20% sucrose (dissolved in reconstitution buffer) in a centrifuge tube. The solutions are centrifuged at 30,000 rpm for 20 hours. Under these conditions, the RNA-protein complex formed during the reconstitution process sediments through the sucrose leaving the excess protein in the supernatant. The resultant pellet can then be resuspended in any desired buffer for subsequent study.

REFERENCES

1. E. Kaltschmidt and H. G. Wittmann, Proc. Natl. Acad. Sci., 67, 1276 (1970).

2. G. R. Craven, P. Voynow, S. J. S. Hardy, and C. G. Kurland, Biochemistry, 8, 2906 (1969).

3. R. R. Traut, H. Delius, C. Ahmad-Zadek, T. A. Bickle, P. Pearson, and A. Tissieres, Cold Spring Harbor Symp. Quant. Biol., 34, 25 (1969).

4. H. G. Wittmann, G. Stöffler, I. Hindennach, C. G. Kurland, L. Randall-Hazelbauer, E. A. Birge, M. Nomura, E. Kaltschmidt, S. Mizushima, R. R. Traut, and T. A. Bickle, Mol. Gen. Genet., 111, 327 (1971).

5. P. Traub and M. Nomura, Proc. Natl. Acad. Sci., 59, 777 (1968).

6. A. Tissieres, J. D. Watson, D. Schlessinger, and B. R. Hollingworth, J. Mol. Biol., 1, 221 (1959).

7. C. G. Kurland, J. Mol. Biol., 18, 90 (1966).

8. P. F. Spahr, J. Mol. Biol., 4, 395 (1962).

9. H. Maruta, T. Tsuchiya, and D. Mizuno, J. Mol. Biol., 61, 123 (1971).

10. J. P. Waller and J. I. Harris, Proc. Natl. Acad. Sci., 47, 18 (1961).

11. S. J. S. Hardy, C. G. Kurland, P. Voynow, and G. Mora, Biochemistry, 8, 2897 (1969).

12. E. Kaltschmidt and H. G. Wittmann, Anal. Biochem., 36, 401 (1970).

13. P. S. Leboy, E. C. Cox, and J. G. Flaks, Proc. Natl. Acad. Sci., 52, 1374 (1964).

14. P. Voynow and C. G. Kurland, Biochemistry, 10, 571 (1970).

15. C. G. Kurland, S. J. S. Hardy, and G. Mora, Methods Enzymol., 20, 381 (1971).

16. P. Traub, S. Mizushima, C. V. Lowry, and M. Nomura, Methods Enzymol., 20, 391 (1971).

17. R. Monier, S. Naomo, D. Hayes, and F. Gros, J. Mol. Biol., 5, 311 (1962).

Chapter 2

PURIFICATION OF INTACT BACTERIAL FLAGELLA AND THEIR COMPONENTS

M. L. DePamphilis*

Department of Biochemistry
University of Wisconsin
Madison, Wisconsin

---

*Present address: Department of Biological Chemistry, Harvard Medical School, Boston, Massachusetts.

INTRODUCTION

Many types of bacteria have long (10-15 μm) helical appendages called flagella that are vital to bacterial motility and are generally considered to be motor organelles. Flagella either are found at one or both poles of a bacterium (polar flagellation) or are randomly distributed over the cell surface (peritrichous flagellation). The number of flagella generally varies from 1 to 15 per cell, but as many as 500 to 5000 per cell may be found on species of Proteus [29].

The intact bacterial flagellum is composed of three structurally defined parts: the filament, the hook, and the basal body [21,32]. The filament is the long helical structure composed of the protein flagellin. The hook is a morphologically distinct unit, generally hook shaped, attached to the proximal end of the filament and terminating in the cell wall. The filament and hook are further distinguished from each other by immunological differences [20,36], by their different solubilities in acid and other reagents [3,34], and by their different thermal stabilities [19]. The basal body is the part of the flagellum attached to the hook and bound into the cell envelope [2,12,16-18,50,54,55]. The hook-basal body complex comprises only 1-2% of the flagellum's length. Iino [30] has recently reviewed the genetics and chemistry of bacterial flagella.

A general structure for the cell envelope of Gram-negative bacteria (exemplified by *Escherichia coli* [48]) consists of three basic components [27]: (1) an inner membrane, the cytoplasmic membrane, made of lipid and protein; (2) an outer membrane, the lipopolysaccharide or L membrane, made of lipopolysaccharide, protein and phospholipid; and (3) an intermediate layer, the peptidoglycan layer, made of peptidoglycan and possibly some protein. The L membrane and peptidoglycan layer are the major components of the cell wall.

The basal body on flagella from *E. coli* and, apparently, other Gram-negative bacteria can be described as four rings mounted on a rod [16]. The ring closest to the hook specifically attaches to the L membrane, and the ring farthest from the hook attaches to the cytoplasmic membrane [17]. The second ring from the hook is apparently attached to the peptidoglycan layer [17]. The site of attachment of the fourth ring, if any, is unclear.

Flagellar basal bodies from *Bacillus subtilis* are similar to those of *E. coli*, except that they have only one pair of rings [17]. The ring farthest from the hook is attached to the cytoplasmic membrane, while the site of attachment of the other ring, if any, is not clear; it is probably anchored into the thick, single-layered cell wall typical of Gram-positive bacteria [27].

Methods are now available for purifying intact flagella with or without outer membrane attached, filaments, flagellin, and hooks. Methods are currently being developed for purifying hook-basal-body complexes.

II. FLAGELLA ASSAYS

The following assays are presently used.

1. Electron microscopy is most commonly used, but is only semiquantitative. Its advantage is that it is the only way to determine the presence and purity of basal bodies. For details, see Chapter 8 on "Negative Staining for Electron Microscopy."

2. To assay purified flagella, the absorption at 280 nm, corrected for light scattering [22], is used after this measurement is related to protein content in the following way. Total nitrogen is measured on purified samples of flagella filaments at different concentrations by a sensitive micro-Kjeldahl procedure [31]. A plot of mg nitrogen x 6.25 against corrected absorbance at 280 nm gives the relationship: 1 A_{280} = 1.85 mg of flagellar protein from _Escherichia coli_ strain AW330 [15].

3. To assay flagella on whole cells and in crude preparations, a radioactive anti-flagella antibody technique was de-

veloped by Grant and Simon [28]. A modification of this method is as follows: Antibody against purified flagella filaments is produced in rabbits; the serum immunoglobulin G was purified on diethylaminoethyl-cellulose [11] and then labeled with ^{125}I, in the presence of Chloramine-T at 5°C [44]. Possibly a more gentle and efficient method for iodination is the use of lactoperoxidase (Calbiochem), as described by Morrison and Bayse [43]. A dilution of labeled antibody is used to prepare a standard curve with known amounts of flagellar filaments. The flagella and antibody are incubated for 30 min at 30°C and then collected on a Gelman cellulose acetate filter, 100-nm pore size, previously soaked in 0.1 M potassium phosphate buffer (pH 7). The excess antibody is washed through with this buffer. Millipore filters cannot be used because they bind the free antibody. The filters are dried under a heat lamp, placed in 10 ml of scintillation fluid [3 g of 2,5-diphenyloxazole (PPO) and 100 mg of 1,4-bis-2-(4-methyl-5-phenyloxazolyl)benzene (POPOP) in 1 liter of toluene] and counted. The range of sensitivity of detection depends on the concentration of labeled antibody used, but it is generally between 0.01 and 0.5 μg of flagellar filaments.

III. PURIFICATION OF INTACT FLAGELLA

A procedure developed by DePamphilis and Adler [15] to purify intact flagella from *E. coli* depends upon solubilizing the cell envelope to release basal bodies, then purifying flagellar filaments under conditions that minimize breakage and yield filaments that retain their hook-basal-body complexes. Lysozyme and ethylene-diaminetetraacetic acid (EDTA) are used to form spheroplasts (osmotically labile bacteria with some cell wall remaining), then the spheroplasts are lysed with a mild, nonionic detergent and the flagella are purified by $(NH_4)_2SO_4$ precipitation, differential centrifugation, and CsCl gradient centrifugation. The same procedure (diagramed in Fig. 1) was also used successfully to purify intact flagella from *B. subtilis*, which has a strikingly different cell wall [27] that is typical of Gram-positive bacteria, and thus, in principle, should be applicable to any bacterium that is senstive to lysozyme.

A similar procedure to purify flagella from *B. subtilis* was developed independently by Dimmitt and Simon [19].

A. Step I: Growth and Harvest of Cells

The *E. coli* cells are grown overnight on a rotary shaker at 35°C in 10 ml of tryptone broth (1% Bacto-Tryptone plus 0.3% NaCl) or glycerol Casamino acids medium [0.2% glycerol 1% Casamino acids, 0.1 M potassium phosphate (pH 7), 0.005 M $(NH_4)_2SO_4$, 0.001 M $MgSO_4$, and 10^{-6} M $Fe_2(SO_4)_3$]. The salts are autoclaved separately from the glycerol and Casamino acids. Thiamine (1 μg/ml) is sterilized with a Millipore filter and included for certain thi$^-$ strains of E. coli. This serves as the inoculum for 1 liter of the same medium in a 6-liter flask that is agitated on a rotary shaker. Just after divergence from exponential growth, at an A_{590} of 1.2 (about 9 x 10^8 cells/ml), 2 liters of cells are harvested by batch centrifugation at 5000 x g for 15 min. Continuous centrifuges, such as the Sharples, should be avoided since they effectively shear off flagellar filaments. The pellets are resuspended in 50 ml of 20% (w/w) sucrose (final volume about 70 ml) by gentle swirling for 1 hr in a rotary shaker that contains ice water. Pellets are suspended gently to prevent loss of flagella from shear.

Tryptone broth or glycerol-Casamino acids medium was superior to minimal medium for the production of highly motile, well-flagellated cells that were suitable for the purification procedure described below.

The quantity of cells was limited by the need to avoid continuous centrifuges, as well as the need to harvest cells before they reach maximum density. Cells in the stationary phase of growth were very sensitive to the lysis procedure, but had fewer flagella per cell. Intact flagella isolated from cells prior to their point of divergence from exponential growth were difficult to purify from material that adhered to the basal bodies (see step V). This appeared to be the result of incomplete degradation in the lysozyme treatment. Osmotic shock lysed fewer of these spheroplasts and yielded larger cell fragments than when spheroplasts were made from cells between exponential and stationary phases of growth.

Finally, although intact flagella could be purified from all strains of E. coli K-12 tested, both the yield and quality of the preparation varied with the strain. This purification procedure was developed with strain AW330 [4]. However, strain KL983 (Hfr) grown on tryptone broth gave a greater yield of purified flagella (S. Larson and J. Adler, personal communication), while flagella were more difficult to purify from several paralyzed mutants derived from strain AW330. One may, therefore, expect various degrees of success with different genera and species. The optimal growth conditions and harvest time that give the most effective cell lysis must be determined for each type of bacterium. The extent of cell lysis (see step II) can be monitored spectrophotometrically [9].

B. Step II: Preparation and Lysis of Spheroplasts

Spheroplasts are formed by the addition of 7 ml of 1 M Tris-HCl (pH 7.8), 2 ml of 0.25% lysozyme in 0.1 M Tris-HCl (pH 7.8) and 0.2 M NaCl, and 6 ml of 0.1 M EDTA in 0.1 M Tris-HCl (pH 7.8) in that order. The suspension is mixed after the addition of each reagent, and the EDTA is added within 30 sec after the lysozyme.

Incubation at 30°C for 1.5 hr with gentle shaking yielded 99% sperhoplasts. Then, addition of 7 ml of 20% (w/w) Triton

X-100 (octylphenoxy polyethoxyethanol, Sigma Chemical Co.) gave a clear viscous lysate in 15 sec. In the lysate, basal bodies had a strong affinity for DNA; the flagella could not be separated from the DNA by sedimentation. Therefore, 0.8 ml of 1 M $MgCl_2$ was added, followed by 1.5 mg of deoxyribonuclease I (Worthington Biochemical Corp.), and the mixture was incubated at 30°C for 20 min. The Mg^{++} must not be added before the Triton X-100, since Mg^{++} prevents solubilization of the outer lipopolysaccharide membrane by the detergent with the result that basal bodies remain encapsulated in this material [13,17].

Lysozyme is purchased as a crystalline enzyme (Worthington); it therefore provides a convenient, rapid method for the degradation of large amounts of the cell walls from a wide variety of bacteria by specifically attacking their peptidoglycan component. In the case of Gram-negative bacteria, the outer membrane must be disrupted before lysozyme can be very effective [8,49]. The EDTA accomplishes this disruption by chelation of divalent metals which results in the release of up to 50% of the lipopolysaccharide material [37,51]. The exact recipe for effective lysozyme treatment should be developed for each bacterial species. One may also consider other cell-wall-degrading enzymes [25,26], penicillin [7] and, with appropriate bacteria, diaminopimelic acid starvation [7].

Tris was the most effective buffer tried for the preparation of spheroplasts with lysozyme and EDTA [49]. In addition, the optimal pH for the purification of intact flagella was between 8.1 and 8.4. At higher pH values, flagella began to fragment. At lower values, material adhered to the basal bodies. All Tris-HCl buffers in this procedure are adjusted to pH 7.8 at 26°C with HCl. The pH will then be 8.2 at 5°C.

Triton X-100 lyses spheroplasts by solubilization of the cytoplasmic membrane; in the presence of EDTA, it also solubilizes the outer membrane [13]. Triton X-100 does not damage flagella, as judged by the fact that motility and growth of E. coli are normal in 1.5% Triton X-100, and no effect on the structure of purified intact flagella is observed after they are treated with 10% Triton X-100.

Brij-58, also a nonionic detergent, was used by Dimmitt and Simon [19] to lyse lysozyme-treated B. subtilis.

Sporulating bacteria such as B. subtilis contain autolytic enzymes that are active when cells are not growing at their maximum rate. The presence of these enzymes and the absence of an outer membrane should simplify the purification of intact flagella from these bacteria.

Some success was achieved in freeing basal bodies from Rhodospirillum [12] by the use of a nonspecific protease [23] to lyse the cells during a 1-2 hr incubation at 35°C, and from Leptospira [46], which have axial filaments, by incubation of cells in 1% deoxycholate for 6 hours at room temperature. These procedures seem more likely to damage the flagellar structure than does the lysozyme-nonionic detergent technique.

C. Step III: Ammonium Sulfate Fractionation

For ammonium sulfate fractionation, the lysate is immediately diluted to 260 ml with cold (5°C) 0.1 M Tris-HCl (pH 7.8; 26°C) containing 5 x 10^{-4} M EDTA (Tris-EDTA buffer) in chilled glassware, and 87 ml of a solution of $(NH_4)_2SO_4$, saturated at 5°C in Tris-EDTA buffer, is rapidly poured into the diluted lysate to give a final solution at 25% of saturation. The suspension is stirred for 2 hr at 5°C. Hopefully, the low temperature, the EDTA, and the precipitation of flagella will stop any adverse effects of the bacterial lysate on the flagellar structure, as the flagella are thus limited to a 20-min exposure to the lysate at 30°C.

When $(NH_4)_2SO_4$ was added slowly at a rate of 0.2 ml/sec with stirring, the final purity and yield were poor. At final concentrations higher than 30% of saturation, flagella were difficult to purify.

Centrifugation of the suspension at 12,000 x *g* for 25 min in an angle-head rotor gives a viscous white material, which floats at the meniscus and adheres to the side of the tube, that contains about 70% of the flagella. This material is collected with a flat spatula. (The use of a swinging-bucket rotor allows all of the material to float at the surface in a dense, easily removed layer.) The fluid is discarded and the tube is rinsed with Tris-EDTA buffer. The rinse is combined with the recovered material to give a final volume of 40 ml.

After 1.0 ml of 20% Triton X-100 is added, the material is gently dispersed. It is then dialyzed twice for 12 hr at 5°C, against 1 liter of Tris-EDTA buffer containing 5 ml of 20% Triton X-100 and 1 ml of toluene (as a bactericide), while it is stirred rapidly enough to create a vortex to disperse the toluene. The initially turbid suspension completely cleared on dialysis. Besides removing low-molecular weight compounds, dialysis allows the EDTA and Triton X-100 to solubilize outer membrane aggregates that formed when Mg^{++} was added. This procedure prevents such membraneous material from adhering to basal bodies.

D. Step IV: Differential Centrifugation

For differential centrifugation, the dialyzed preparation is diluted to 50 ml with Tris-EDTA buffer. In each of two Spinco SW-25 tubes, 25 ml of the diluted material is placed on top of 2 ml of 20% (w/w) sucrose that had been layered over 3 ml of 60% sucrose. The sucrose solutions are prepared in Tris-EDTA buffer. The sucrose cushion is necessary when intact flagella are sedimented; otherwise, tightly packed pellets form that are difficult to resuspend without breaking off about 70% of the basal bodies. After centrifugation in a Spinco SW-25 rotor at 20,000 rpm for 1 hr, the liquid above the sucrose cushion (fraction B) is removed with a pipette and discarded, since further treatment of fraction B (see step V) does not yield pure flagella. To the sucrose cushion 5 ml of Tris-EDTA buffer is added. After it is gently mixed, the sucrose is removed by dialysis overnight against Tris-EDTA buffer. The suspension is then diluted to 25 ml with the buffer, and 0.15 ml of 20% Triton X-100 is dissolved in the suspension. This solution is centrifuged at 4000 x *g* for 10 min; the pellet, containing unlysed cells and large aggregates, is discarded, while the supernatant material, containing about 40% of the flagella (fraction A), is used in the next step.

E. Step V: CsCl Gradient Centrifugation

For CsCl gradient centrifugation, fraction A is diluted to 27 ml with Tris-EDTA buffer, and 12.1 g of CsCl is added in one portion and rapidly dissolved. After centrifugation in a Spinco SW-25 rotor at 22,000 rpm for 50 hr, the tube contains a single translucent band in the center of the gradient, at a density of 1.30 g/ml, that contains about 98% of the flagella in the tube, or about 40% of the original amount of flagella. However, sometimes fraction A gives, in addition, a white band at a lighter density that contains as much as half of the flagella in the tube, and is contaminated with carbohydrate-containing material. Electron microscopy of this band shows that the basal bodies are embedded in cell material. The investigator can fractionate the gradient either by first removing the viscous oily material at the top with a pipette and then siphoning from the bottom, or by use of a device for puncturing the bottom of the tube with a syringe needle [45]. The fraction containing the translucent band is then dialyzed against Tris-EDTA buffer to remove the CsCl.

The appearance and number of bands on CsCl, and the purity and quantity of flagella recovered, were markedly influenced by the rate of addition of CsCl to the suspension of flagella and by the age of the cells at harvesting. When the CsCl was dissolved in 0.5-g quantities, a wide and continuous density distribution of material resulted, with a faint, poorly resolved flagella band. Addition of all the CsCl at once to an identical, replicate sample gave a single visible band containing four times as much purified flagella. Under the above growth conditions, cells departed from exponential growth on tryptone at an A_{590} of 1.0. Cells harvested at an earlier time (A_{590} of 0.75) did not yield pure, well-defined bands of flagella in a CsCl gradient.

The major value of the CsCl gradient is that it separates purified intact flagella from intact flagella with cell wall or cytoplasmic material adhering to the basal bodies, while simultaneously concentrating the flagella without damage due to shear forces and, in addition, removes the Triton X-100 which concentrates at the meniscus.

IV. COMMENTS

The flagella obtained by the preceding methods are predominantly "intact" as filament-hook-basal-body complexes with a well-defined structure [16,19]. The remaining material is pieces of filaments that fragmented during the purification. The preparations are free of nonflagellar material that would be visible in an electron microscope. Less than 1% of the protein has a molecular weight of 10^5 or less, as judged by polyacrylamide-gel electrophoresis. Based on the flagellar antibody assay, filaments comprise 98% of the protein. Carbohydrates never exceeded 2.2% and inorganic phosphate was never greater than 0.15%. Ultraviolet absorption spectra were identical to those from purified flagellar filaments (see below), and typical of purified proteins with a ratio of absorbance at 280/260 of 1.35 to 1.40. Purified intact flagella had the same density in CsCl as purified flagellar filaments, and this density was not altered by the presence of 0.5% Triton X-100. By these criteria of purity, only minor amounts of nucleic acids, phospholipids or detergent, carbohydrates, and proteins were present. The only detectable components were flagella and, after appropriate depolymerization, flagellin. However, these data would not reveal the composition of the hook or basal body since these comprise at most 2% of the preparation.

In the event that the above procedures do not yield pure flagella, one may find that repeating step V or IV will complete the purification. In addition, Gelman cellulose acetate filters, pore size 100 to 200 Mm, are useful for both the concentration of small volumes of flagella and further purification. Resuspension of flagella from filters must be done slowly, by soaking the filter in a buffer, in order to prevent fragmentation of filaments and basal ends.

The basal bodies of purified intact flagella from *E. coli* and, to a much lesser extent, *B. subtilis* tended to aggregate. This phenomenon became extensive when concentrated solutions of flagella (0.8 mg/ml) were stored at 5°C for one week. The flagella could not be disaggregated by extensive dialysis against Tris-EDTA buffer or the same buffer containing 2.0 M KCl. The presence of 0.1% Triton X-100 caused partial disaggregation, but the best remedy was to store flagella in dilute solutions of 0.1 mg/ml or less. Aggregation was caused by the extraction of the Triton X-100, which floats on the CsCl gradient, and the simultaneous concentration of flagella. No aggregation of basal bodies was observed before the CsCl step as long as Triton X-100 was present.

A critical question is whether a vital component was lost when the "intact" flagellum was isolated. Until the flagellum can function in an *in vitro* system, one cannot prove it was prepared in its complete form. To the best of my knowledge, all methods available for producing osmotically labile bacteria, whether they use enzymes, antibiotics, or metabolic restraints to disrupt the cell wall, result in flagellated but nonmotile cells as soon as cell-wall rigidity is lost. Therefore, the possibility remains that a fragile or unattached component of the flagellar base is missing in these preparations.

V. PURIFICATION OF INTACT FLAGELLA WITH FRAGMENTS OF OUTER (L) MEMBRANE ATTACHED

Fragments of L membrane in the form of vesicles are found attached to the basal bodies of intact flagella when the purification procedure described for intact flagella is modified in one important aspect. Once spheroplasts have been formed, Mg^{++} is used instead of EDTA in order to prevent the L membrane from being solubilized by Triton X-100 [13,17].

After cells are incubated for 1 hr with lysozyme and EDTA, 1 M $MgCl_2$ is added to give a molar ratio of Mg^{++} to EDTA of 6, and the suspension is incubated at 30°C for 30 min. Lysis occurs within one minute after the addition of Triton X-100. The lysate is incubated with deoxyribonuclease I. The lysate is turbid because of the presence of L membrane.

The buffer used throughout the remainder of the procedure is 0.1 M Tris and 10 mM $MgCl_2$ adjusted to pH 7.8 with HCl at 26°C (Tris-Mg buffer). The $(NH_4)_2SO_4$ precipitation yields a dense white pellet and a viscous floating layer that are combined and dialyzed against Tris-Mg buffer containing 0.1% Triton X-100 and toluene, 1 ml/liter. After dialysis, the material is placed over a sucrose cushion and centrifuged in a Spinco SW-25 rotor at 10,000 rpm for 1 hr at 5°C. The sucrose cushion is collected, dialyzed against Tris-Mg buffer, and diluted with buffer, then Triton X-100 is added to give a final concentration of 0.1%. The material is centrifuged for 10 min at 5000 x g (R_{max}); the pellet is discarded, and the supernatant is fractionated on a CsCl gradient as previously described.

The result is a single white flocculent band, 50-55 mm from the meniscus, at a density of 1.34 g/ml, that is identified as L membrane with flagella attached; the identification was based on negative staining, thin sectioning, chemical composition, and comparison with purified lipopolysaccharide from the same strain of *E. coli* [13,17], as well as on the basis of attachment of T2 and T4 phage [14,17]. The supernatant liquid from the centrifugation at 10,000 rpm yields several band at densities between 1.32 and 1.30 g/ml that contain predominantly flagella with all basal bodies attached to an L-membrane vesicle. Many of these vesicles are uniform in size and appearance. The vesicles are always found attached specifically to the ring closest to the hook.

VI. PURIFICATION OF FLAGELLAR FILAMENTS

Flagellar filaments (often referred to in the literature as flagella or sheared flagella) are easily removed from bacteria by shearing [47,53]. Either a blender or a commercial paint shaker is employed. The resulting flagellar filaments are then purified by differential centrifugation [1,33] or ion-exchange chromatography [40]. The following procedure is based on these methods.

Motile bacteria are harvested from their culture medium in late-exponential growth by centrifugation and resuspended in either 0.1 M phosphate, pH 7.0, or 0.1 M Tris, pH 7.8, to a density of approximately 10^{10} cells/ml. Flagellar filaments are sheared off at 4°C in a Waring blender (or Lourdes homogenizer [40] or Omni-Mixer [33]) at the lowest speed necessary to deflagellate within 2 min. The *E. coli* flagellated K-12 species lost 99% of their motility with no loss of viability after being blended for 45 sec at 19,500 rpm. Efficient, rapid shearing depends upon a high cell density. The remaining steps are carried out at 4°C.

The suspension is diluted sixfold in buffer and subjected to two cycles of differential centrifugation at 12,000 x *g* for 20 min to sediment cells and cell debris, followed by centrifugation at 55,000 x *g* for 1 hr to sediment flagellar filaments. If the original suspension is too concentrated, poor separation may result during centrifugation. Pure filament pellets look like clear gelatin. Contaminating cell debris looks white or brown. Pellets are soaked overnight in buffer before gentle resuspension to avoid unnecessary fragmentation of filaments.

If a piliated bacterium was used, the resuspended material may contain trace amounts of pili even though pili do not sediment well under the above conditions since they are about half of the diameter and one-fifth of the length of the filament pieces. So far only Gram-negative bacteria have been found with pili [24]. These include _E. coli_, _Proteus_, _Salmonella_, and _Pseudomonas_. Pili can be detected on bacteria with an electron microscope after shadowing [10] or negative staining [24]. I have found that banding flagellar filaments in CsCl removes all traces of pili (see step V on CsCl gradient centrifugation). In addition, a blender speed can be chosen that removes most of the flagellar filaments but little of the pili [47].

Criteria for purity are generally electron microscopy, polyacrylamide gel electrophoresis [6,15,41], and the absence of inorganic phosphate and carbohydrate. Since normal flagellar filaments are too large to enter the gel, the absence of acidic or basic protein bands before depolymerization is a good criterion of purity. Normally, only one band is seen on 7% gels after depolymerization, but some strains of _Salmonella_ [30] and a strain of _Bacilius pumilis_ [35] contain two different flagellins. Hooks are not easily dissociated and thus will not enter the gel [35].

Flagellar hooks may be considered a significant contaminant in some preparations. At present, the only way to remove hooks from filaments is to dissociate the filament sediment hooks away from flagellin, and then allow the flagellin to polymerize [30; also see Section VII on purification of flagellin]. Reconstituted flagellar filaments are indistinguishable from native filaments [30].

VII. PURIFICATION OF FLAGELLIN

Flagellar filaments are dissociated by various agents that include acid, base, heat, urea, guanidine, and sodium dodecyl sulfate [3,34]. Flagellin is frequently prepared from purified flagellar filaments that are treated with either heat [5] or 0.1 M HCl [1] and then centrifuged at 100,000 x *g* for 1 hr at 5°C; any pellet is discarded. Minimal conditions for complete filament dissociation are conveniently determined by measurement of the viscosity after various treatments with an Oswald-type viscometer that has a flow time of 25 to 75 sec for water. As the pH of a filament suspension (0.5 mg/ml in 0.05 M KCl at 30°C) is lowered below pH 7, its viscosity increases; the viscosity reaches a maximum near pH 4 and then drops sharply near pH 3.5. The exact values will vary with ionic strength, temperature, and source of flagella. When flagellar filaments are heated to different temperatures, a melting curve is observed that spans approximately 8°-10°C and has a midpoint generally near 60°C, but that may be as low as 46°C. The midpoint increases with ionic strength and is characteristic for flagella from different organisms [18,34,42]. Dissociation proceeds more rapidly if filaments are first fragmented by sonication [33] or passage through a 24-gauge needle about 10 times [6,18].

Flagellin from *Proteus vulgaris* [38], *Bacillus pumilus* [1], *B. stearothermophilus* [1], and *B. subtilis* [41], prepared with acid or base, spontaneously reaggregates into filaments as the pH is returned to 7. Reaggregation is prevented in a 2-mg/ml solution at 25°C if 0.15 M phosphate is present [1]. Otherwise, reaggregation is retarded by low temperatures and high-salt concentrations and low-protein concentrations (less than 0.2 mg/ml) [1]. In contrast, *Salmonella* flagellin prepared (after sonication) by heating at 65°C for 5 min in 0.05 M phosphate, pH 7 is stable; the ends of added filament fragments must act as nuclei for polymerization to occur [5,6].

VIII. PURIFICATION OF FLAGELLAR HOOKS

Methods for the isolation of hooks are based on the selective dissociation of filaments to flagellin under conditions where hooks remain intact. Dimmitt et al. [18] begin with purified intact flagella and their method, described in the next section, yields hook-basal-body complexes. Koffler et al. [35] and Abram et al. [3] begin with isolated flagellar filaments from *B. pumilus* that retain up to 40% of the hooks [35].

Filaments (about 1 mg/ml) are fragmented by freeze-thawing [35] and then adjusted to pH 3 with 0.1 M HCl and incubated for 30 min at 37°C. Hooks are stable under these conditions, but will disintegrate at pH 1-2 [3,16]. Hooks and contaminating cell debris can be centrifuged at 5000 x *g* for 15 min at 4°C. The precipitate from the acid-treated filament preparation is very difficult to resuspend, probably because of aggregated denatured cell-envelope material. However, addition to the pellet of one-fifth of its original volume of 4% Triton X-100 solubilizes most of the nonhook material in 6 hr at 37°C. Free hooks can then be sedimented at 100,000 x *g* for 1 hr and further purified on a renografin gradient (1.16-1.34 g/ml Renografin-76, methyl glucamine diatrizoate, Squibb, Inc., New York, N.Y., in 0.1 M Tris-HCl, pH 8). Hooks are found at a density of 1.20 g/ml after centrifugation for 8 hr at 200,000 x *g* and 15°C. Purified hooks constitute about 0.5% by weight of the original filament preparation [35].

In an alternative procedure, Abram et al. [3] find that rapid stirring of the filaments in 50% ethanol, 10^{-4} M HCl at 4°C for 9 hr disintegrates filaments into fine strands that can be sedimented at 20,000 x *g* and 4°C for 1 hr and then resuspended to their original volume in 10^{-3} M HCl. After rapid mixing for 9 hr at 4°C, flagellin strands completely dissolve and hooks, with and without membrane material attached, are sedimented at 100,000 x g for 2 hr. Most of the membrane material is solubilized by resuspension of the pellet to 2% of its original volume with either deoxycholate (3% w/v, pH 9.0) or saponin (0.3% w/v, pH 7.0) and incubation for 2 hr at 37°C. A second sedimentation at 100,000 x g for 2 hr precipitates hooks, which comprise about 0.2% by weight of the original flagella suspension. However, they are only about 50% pure, and contaminants are visible in the electron microscope; when the solution is adjusted to pH 4-5, the hooks precipitate along with the contaminants. Sedimentation at 3000 x g for 10 min leaves about 5% of the material, as highly purified hooks, in the supernatant.

The purity of hook preparations was determined by electron microscopy [3,19], sedimentation in sucrose gradients [18], and the absence of flagellin detectable by polyacrylamide gel electrophoresis [18,35]. Two difficulties with the above procedures are: (1) the yields are small and, therefore, hooks must be radioactively labeled by growth of the bacteria with an appropriate isotope in the medium before any chemical and physical studies can be done [18,35]; (2) it is not clear whether these methods will be applicable to Gram-negative bacteria such as *E. coli* and *Salmonella* since the amount of hooks found in filament preparations from these organisms may be much smaller.

IX. PURIFICATION OF HOOK-BASAL-BODY COMPLEXES

The same principles used for the isolation of hooks from flagellar filaments are successful for the isolation of hook-basal-body complexes from intact flagella [16,19]. Viscometry or flagellar antibody assays [16,19] can be used to determine the minimal conditions for dissociation of filaments. Intact flagella can be more resistant to dissociation than flagellar filaments without hooks attached [19].

Bacillus subtilis is grown in minimal medium supplemented with 0.08 mCi/ml of ^{3}H-labeled leucine and histidine [18]. Intact flagella are isolated [19] and sheared by passage through a 24-gauge needle 10 times to make them more susceptable to dissociation. The sample is heated for 15 min at 65°C to dissociate the filaments. With this particular strain, heating at 56°C for 15 min destroys 50% of the antigenic activity of the filaments [19]. The solution is layered on a gradient composed of three layers of sucrose: 38%, 45%, and 53% prepared in 0.01 M Tris, pH 8.0, and 0.5% Brij-58 (TB buffer). After centrifugation for 5 hr at 100,000 x *g* a band is present at the interface between the 38% and 45% sucrose that contains hooks, hook-basal-body complexes, and hook-basal-body complexes attached to membrane fragments. Most of the flagellin is removed at this step. The band is collected and dialyzed against TB buffer to remove the sucrose, then further purified by isopycnic centrifugation in Renografin-76. The sample is layered on a linear (1.14-1.25 g/ml) preformed Renografin gradient in TB buffer. Centrifugation at 144,000 x g for 12 hr separates the particles into hooks at 1.20-1.22 g/ml, hook-basal-body complexes at 1.18-1.20 g/ml, and hook-basal-body complexes attached to membrane at 1.16 g/ml and less.

A second approach is being developed for the purification of hook-basal-body complexes from E. coli (S. Larsen and J. Adler, personal communication). Since basal bodies from E. coli become degraded at pH values less than 3.0 [16], urea is used to dissociate the filaments. A concentrated solution of urea is added to a sample of purified, intact flagella to give a 1-2 ml volume that contains 5 M urea. The sample is incubated for 30 min at 25°C and then placed on top of a 0.9 x 18 cm agarose column (Bio Rad A-5M agarose) at 5°C to remove the flagellin. The hook-basal-body complexes elute in the void volume with 0.1 M Tris, 5 x 10^{-4} M EDTA, pH 7.8 at 25°C. Electron microscopy shows that 99% of the filaments are removed by this procedure. Repetition of the procedure with the hook-basal-body complexes gives a preparation containing no detectable filaments.

A third possible method could be to shear away flagellar filaments from the cells and then to isolate the outer membrane with basal bodies still firmly attached [17]. The purified outer membrane could then be solubilized to release hook-basal-body complexes and isolated basal bodies [13]. The flagellar components could then be separated from solubilized outer membrane by gel filtration or isopycnic centrifugation. The advantage of this approach is that large quantities of cells could be harvested with a Sharples continuous centrifuge, which is very effective at shearing off flagellar filaments. This may be a solution to the problem of purifying large amounts of basal bodies for chemical and physical studies.

Proteus may be a good source of basal bodies, since the density of flagella on these cells is much greater than that of other bacteria [29]. However, *Proteus* species are generally difficult to convert into spheroplasts.

Finally, hook-basal-body complexes can be further degraded at low pH [3,16], or by heating 3 min at 100°C in 1% sodium dodecyl sulfate, 1% β-mercaptoethanol, and 0.01 M Tris, pH 9.1 [18]. Perhaps more specific breaks can be made by the use of proteolytic or lipolytic enzymes.

REFERENCES

1. D. Abram and H. Koffler, *J. Mol. Biol.*, *9*, 168 (1964).

2. D. Abram, H. Koffler, and A. E. Vatter, *J. Bacteriol.*, *90*, 1337 (1965).

3. D. Abram, J. R. Mitchen, H. Koffler, and A. E. Vatter, *J. Bacteriol.*, *101*, 250 (1970).

4. J. B. Armstrong and J. Adler, *Genetics*, *56*, 363 (1967).

5. S. Asakura, J. Mol. Biol., 10, 42 (1964).

6. S. Asakura, G. Eguchi, and T. Iino, J. Mol. Biol., 16, 302 (1966).

7. M. E. Bayer, J. Gen. Microbiol., 46, 237 (1967).

8. D. C. Birdsell and E. H. Cota-Robles, J. Bacteriol., 93, 427 (1967).

9. D. C. Birdsell and E. H. Cota-Robles, Biochem. Biophys. Res. Commun., 31, 438 (1968).

10. C. C. Brinton, Trans. N. Y. Acad. Sci., 27, 1003 (1965).

11. D. H. Campbell, Methods in Immunology; a Laboratory Text for Instruction and Research, Benjamin, New York, 1964, p. 122.

12. G. Cohen-Bazire and J. London, J. Bacteriol., 94, 458 (1967).

13. M. L. DePamphilis, J. Bacteriol., 105, 1184 (1971).

14. M. L. DePamphilis, J. Virology, 7, 683 (1971).

15. M. L. DePamphilis and J. Adler, J. Bacteriol., 105, 376 (1971).

16. M. L. DePamphilis and J. Adler, J. Bacteriol., 105, 384 (1971).

17. M. L. DePamphilis and J. Adler, J. Bacteriol., 105, 396 (1971).

18. K. Dimmitt, S. Emerson, K. Tokuyasu, and M. Simon, in Behavior of Microorganisms, Symposia of the Tenth International Congress for Microbiology, Mexico City, 1970, in press.

19. K. Dimmitt and M. Simon, J. Bacteriol., 105, 369 (1971).

20. K. Dimmitt and M. Simon, Infec. Immun., 1, 212 (1970).

21. R. N. Doetsch and C. J. Hageage, Biol. Rev., 43, 317 (1968).

22. S. W. Englander and H. T. Epstein, Arch. Biochem. Biophys., 68, 144 (1957).

23. J. C. Ensign and R. S. Wolfe, J. Bacteriol., 91, 524 (1966).

24. J. A. Fuerst and A. C. Hayward, J. Gen. Microbiol., 58, 227 (1969).

25. J. M. Ghuysen, Bacteriol. Rev., 32, 425 (1968).

26. J. M. Ghuysen, J. L. Strominger, and D. J. Tipper, in Comprehensive Biochemistry (M. Florkin and E. H. Stotz eds.), Elsevier, Amsterdam, 26A, 1968, p. 53.

27. A. M. Glauert and M. J. Thornley, Ann. Rev. Microbiol., 23, 159 (1969).

28. G. F. Grant and M. Simon, J. Bacteriol., 95, 81 (1968).

29. J. Hoeniger, J. Gen. Microbiol., 40, 29 (1965).

30. T. Iino, Bacteriol. Rev., 33, 454 (1969).

31. M. J. Johnson, J. Biol. Chem., 137, 575 (1941).

32. T. M. Joys, Antonie van Leeuwenhoek J. Microbiol. Serol., 34, 205 (1968).

33. D. Kerridge, R. W. Horne, and A. M. Glauert, J. Mol. Biol., 4, 227 (1962).

34. H. Koffler, Bacteriol. Rev., 21, 227 (1957).

35. H. Koffler, R. W. Smith, J. R. Mitchen, and E. McGroarty, in Behavior of Microorganisms, Symposia of the Tenth International Congress for Microbiology, Mexico City, 1970, in press.

36. A. M. Lawn, Nature, 214, 1151 (1967).

37. L. Leive, V. K. Shovlin, and S. E. Mergenhagen, J. Biol. Chem., 243, 6384 (1968).

38. J. Lowy and M. W. McDonough, Nature, 204, 125 (1964).

39. H. H. Martin, J. Theoret. Biol., 5, 1 (1963).

40. R. J. Martinez, J. Gen. Microbiol., 33, 115 (1963).

41. R. J. Martinez, A. T. Ichiki, N. P. Lundh, and S. R. Tronick, J. Mol. Biol., 34, 559 (1968).

42. R. J. Martinez and E. Rosenberg, J. Mol. Biol., 8, 702 (1964).

43. M. Morrison and G. S. Bayse, Biochemistry, 9, 2995 (1970).

44. P. J. McConahey and F. J. Dixon, Intern. Arch. Allergy, 29, 185 (1966).

45. E. H. McConkey in Methods in Enzymology (L. Grossman and K. Moldave, eds.), Academic Press, New York, 12A, 1967, p. 620.

46. R. K. Nauman, S. C. Holt, and C. D. Cox, J. Bacteriol., 98, 264 (1969).

47. C. Novotny, J. Carnahan, and C. C. Brinton, J. Bacteriol., 98, 1294 (1969).

48. S. dePetris, J. Ultrastruct. Res., 19, 45 (1967).

49. R. Repaske, Biochem. Biophys. Acta, 30, 225 (1958).

50. A. E. Ritchie, R. F. Keller, and J. H. Bryner, J. Gen. Microbiol., 43, 427 (1966).

51. S. W. Rogers, H. E. Gilleland, and R. G. Eagon, Can. J. Microbiol., 15, 743 (1969).

52. J. Stensch and H. Koffler, Federation Proc., 21, 406 (1962).

53. B. A. D. Stocker and J. C. Campbell, J. Gen. Microbiol., 20, 670 (1959).

54. W. van Iterson, J. F. M. Hoeniger, and E. N. van Zanten, J. Cell Biol., 31, 585 (1966).

55. Z. Vaituzis and R. N. Doetsch, J. Bacteriol., 100, 512 (1969).

NOTE ADDED IN PROOF

Flagellar hooks were recently purified from *Salmonella* by H. Kagawa, S. Asakura, and T. Iino, Journal of Bacteriology 113: 1474 (1973).

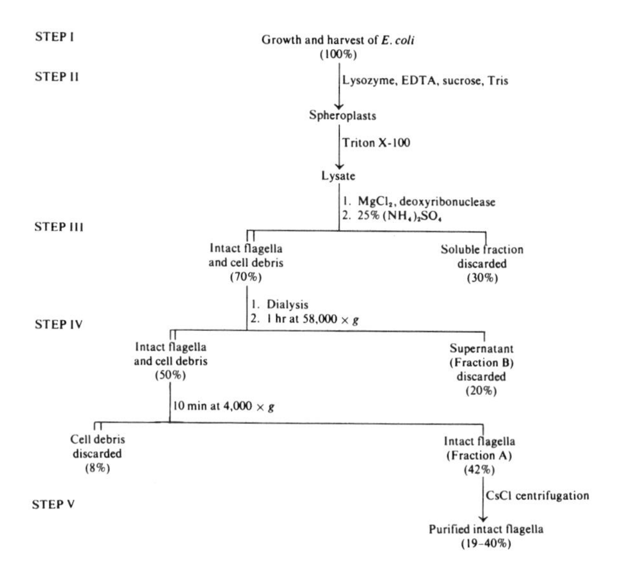

FIG. 1. Flow sheet for the purification of intact flagella from E. coli. Recoveries are shown in parentheses. The original amount of flagella, 2.4 mg/10^{12} cells, is defined as 100%.

Chapter 3

ISOLATION OF PLASMA MEMBRANES FOR CELL SURFACE MEMBRANE RECEPTOR STUDIES

David M. Neville, Jr. and C. Ronald Kahn

Section on Biophysical Chemistry
Laboratory of Neurochemistry
National Institutes of Health
Bethesda, Maryland

and

Diabetes Section
Clinical Endocrinology Branch
National Institute of Arthritis, Metabolism and Digestive Diseases
National Institutes of Health
Bethesda, Maryland

I. INTRODUCTION

It has been twelve years since the demonstration that purified fractions of plasma membranes can be isolated from nucleated mammalian cells [1]. During this time steady progress has been made in characterizing the large number of enzymatic, protein, glycoprotein, and lipid components of these complex structures [2-5].

At the same time that the basic knowledge of membrane components was increasing, more and more investigators became aware of the crucial importance of the plasma membrane in regulating cellular metabolism, differentiation, and growth. The surface membrane separates the external environment from the internal environment and, therefore, is likely to be adapted to receiving signals from the external environment which will modify the intracellular environment. This concept has been held for many years by those interested in membranes [6-9]; however, the early evidence was often circumstantial, and experiments to test the hypothesis were inconclusive or lacked the necessary specificity in formulation and methology. The events which have changed this state of affairs are (1) the discovery of cyclic AMP as a regulator of intracellular enzyme activity, (2) the localization of hormone-stimulated adenyl cyclase to the plasma membrane of many cells [10-12], and (3) the elucidation of polypeptide hormone receptors on cell surface membranes by direct study with biologically active labeled hormone [13-18].

Now that cell surface receptors have become a quantifiable biochemical reality, we may anticipate that the next step will be to identify those factors that control membrane receptor concentration and to discover the role that these receptors play in pathological states. A fruitful approach to this problem will be the study of membrane receptors in animal diseases that result from the inheritance of a single genetic defect. Another approach will be the study of membrane receptors from cell lines that have been transformed by oncogenic viral agents. Both types of study pose particular problems for the investigator preparing plasma membranes. It is the purpose of this chapter to consider some of these problems and to present the available methods for minimizing them.

II. PROBLEMS IN COMPARING MEMBRANES FROM NORMAL OR PATHOLOGICAL STATES

In assessing the role of a specific membrane receptor in a pathological state, one wishes to quantitate the receptor concentration on the membrane so that a quantitative relationship can be determined between the receptor concentration and the physiologic response that is stimulated. The major problem posed in this type of study is how to accurately determine the purity of the plasma membrane preparation. The purity must be known in order that the concentration of plasma membrane can be known, and because cytoplasmic contaminants may also contain receptors that may be of different affinity or specificity or concerned with different functions. This problem is complicated by the following considerations. The plasma membrane derived from cells adherent to other cells or surfaces is heterogeneous in structure and may be heterogeneous in receptor concentration. For example, the plasma membrane of each hepatocyte exists in three morphologies [1]: a single membrane exhibiting small protrusions and invaginations bordering on the sinusoid known as the blood front; a membrane closely opposed to the membrane of the neighboring liver cells exhibiting numerous desmosomes known as the cell front; and a rich protrusion of long villous processes forming the bile canaliculus known as the bile front. Because most membrane isolation procedures achieve recoveries of less than 25 percent, each isolation procedure is potentially a selection procedure. In addition, membranes derived from organs may contain contributions from several different cell types. The problems of selection become particularly worrisome when one is comparing a pathological state with a normal state, because often the gross structure of the organ or cell is altered by the differences between the two states. For example, in the obese hyperglycemic mouse (C57BL/6J ob/ob), the liver is grossly fatty and has a volume and weight almost double that seen in the phenotypically normal thin litter mates [19]. Similarly, tissue-cultured cell lines transformed by virus have altered size, shape, and cellular fragility. The result is that during membrane isolation, a selection process is being imposed on different starting materials, and the purified membrane products may not be the same in each case. Finally, one cannot assume that the ratio of plasma membrane to cytoplasmic membrane contaminants is the same in the pathological case and its normal control, nor can one assume that the ratio of the differentiated parts of the plasma membrane is the same in the normal and the pathologic states. For these reasons, particular attention must be paid to the quantitation of yield and purity of purified plasma membranes.

III. BIOCHEMICAL QUANTITATION OF YIELD AND PURITY

A. Assessing Cytoplasmic Contamination

Biochemical quantitation of yield and purity of a membrane fraction depend on knowledge of the organelle composition of the initial homogenate and knowledge of marker enzymes that are specifically associated with the various organelles. This type of information can be obtained by fractionation of homogenates with respect to sedimentation rates and density, while a large number of different enzyme-specific activities is followed. When this is done, it is noted that groups of enzymes fractionate together. The various groups identify different organelles. If a group of enzymes is completely separated from other groups, the organelle is considered to be purified. The specific activity of an enzyme in such a group is the specific activity of the enzyme in the purified organelle. The contribution of organelle protein to homogenate protein can then be determined by

$$\underline{E}_{org} \quad \frac{U}{\text{mg org. prot.}} = \underline{E}_{hom} \quad \frac{U}{\text{mg hom. prot.}} \times \frac{\text{hom. prot. mg/ml}}{\text{org. prot. mg/ml}} \qquad (1)$$

where $\underline{E}_{org}$ is the specific activity of an organelle enzyme in enzyme units per mg of organelle protein, and $\underline{E}_{hom}$ is the specific activity of the organelle enzyme in the homogenate in units per mg of homogenate protein. Often enzyme groups are not completely separated and corrections for cross contamination must be applied. This type of procedure was introduced by de Duve and co-workers and led to the identification of the lysosome as a discrete organelle [20]. The procedure has been applied to liver [21] and kidney [22]. Once the right- or left-hand side of Eq. (1) is known, the contamination of any fraction by the organelle can be calculated by

$$\frac{\underline{E}_{Fx} \quad \frac{U}{\text{mg Fx prot.}}}{\underline{E}_{org} \quad \frac{U}{\text{mg org. prot.}}} = \frac{\text{org. prot. mg}}{\text{Fx prot. mg}} \tag{2}$$

where $\underline{E}_{Fx}$ is the specific activity of an organelle enzyme per mg of the fraction protein. The calculation assumes, in addition to strict localization of marker enzymes to a single organelle type, that enzyme specific activities are unchanged by the fractionation procedures. The contamination rates reported here for the purification steps of mouse-liver-plasma membranes were calculated by this method using de Duve's values derived from rat liver on organelle composition. As a mitochondrial marker enzyme we use succinate cytochrome C reductase, since this activity is present in both the inner and outer mitochondrial membranes [23], which could be dissociated by the hypotonic isolation media. Care must be taken to use the appropriate inhibitors and control blanks when measuring phosphate-splitting enzymes because of the broad specificity of many phosphatases. Glucose-6-phosphatase, a microsomal marker, should be measured in the presence of both EDTA and potassium fluoride to inhibit alkaline and acid phosphatase as described by Hubscher and West [24]. Since alkaline phosphatase can cleave nucleotide phosphates, measurement of the plasma membrane marker 5'-nucleotidase in the presence of alkaline phosphatase requires a blank reaction with either 2'- or 3'-AMP [25]. In order to minimize errors in enzyme specific-activity values, certain precautions should be taken. If Lowry or biuret protein determinations are used, dilutions of homogenate and dilutions of the protein standard should be compared as to color production, and determinations should be performed only where both curves are linear (for liver fractions below 0.45 absorbance units/cm at 750 mμ for Lowry, with bovine serum albumin taken as standard). Enzyme assays should be checked for linearity with respect to enzyme units per assay and protein concentration. If total marker enzyme activity varies during fractionation, controls that use a mix of homogenate and purified fractions may be necessary. Many procedures influence enzyme activities, such as freezing and thawing, storage at 4°C, and dilution, to mention a few. When comparing different fractions, make certain they each have had the identical previous history.

The de Duve type of fractionation spectra is a valuable method for approximating the composition of membrane fractions.

However, the organelle composition and the marker enzymes derived for liver and kidney cannot be applied indiscriminately to other tissues. Many tissues have organelles unique for that tissue. In addition, glucose-6-phosphatase values are not a useful microsomal marker, except in liver and kidney, since in other tissues the values are exceedingly low [26]. The lack of organelle composition data and marker enzyme spectra on other tissues and cultured cell lines is a major handicap in assessing the purity of cell-membrane preparations which have been reported from these sources. However, the following references provide some enzyme or other marker data for plasma-membrane fractions isolated from intestinal brush border [27,28], pig lymphocyte [29], L cells [30], chick embryo fibroblasts [31], HeLa cells [32], cardiac muscle [33], rat brain [34], and rat myometrium [35].

B. Enrichment of Plasma Membranes Assessed by Marker Enzymes

The theoretical yield of plasma membranes can be calculated from geometrical considerations. The isolated mouse liver cell (thin mouse) is approximately spherical and has an average diameter of 26×10^{-4} cm [36]. From the width of the plasma membrane, 80×10^{-8} cm [2], the membrane protein-to-lipid ratio, 0.47 [37], the protein content per cell, 15%, and the density of the cell, 1.2, we calculate a ratio of membrane protein to homogenate protein of 0.51×10^{-2} (assuming that all homogenate protein is derived from hepatocytes). The yield for thin mice is 0.37×10^{-2}, which would correspond to a 90% recovery at each step. A more reasonable figure would be 60-80% recovery per step putting the over-all expected yield between 20 and 50%. Our recovery appears high and could reflect (1) undetected contaminants, (2) adsorption of homogenate protein to membranes, or (3) more surface membrane than that given by a spherical model, owing to invaginations and protrusions. A 2- to 5-fold increase in area seems reasonable. Beyond this, very special and obvious morphologies are required. (The brush border of the renal proximal tubule surface area is increased 40-fold [38].)

When performing expected yield calculations based on geometry, all the cell types within an organ must be considered. In the liver 30% of the cells are littoral cells although these occupy only 10% of the volume. Their potential contribution of surface membrane can be assessed by calculation of the external sinusoidal capillary surface area and multiplication by 2. This calculation is simplified by the knowledge that about one-third of each hepatocyte surface faces a sinusoid [39]. Sinusoidal membranes could contribute as much as $2 \times 1/3 / [1 + (2 \times 1/3)]$ or 40% of the total surface membrane. This contribution would not be detected by general membrane marker enzymes. Similar considerations apply for any highly vascularized tissue.

The ratio of plasma membrane protein to homogenate protein

can also be calculated by Eq. (1) if one assumes that the purified membrane fraction represents a pure organelle. Since the total observed contamination is less than 10%, the assumption appears justified. This calculation reduces to taking the reciprocal of the purification factor of the marker enzyme, in this case 5'-nucleotidase, giving a ratio of membrane protein to homogenate protein of 5×10^{-2}. Similar purification factors for this enzyme have been reported by others [2,37]. Touster and co-workers have found that if, in addition to sheetlike membranes, a small vesicular fraction rich in 5'-nucleotidase is isolated from liver, both fractions represent 1.6% of the homogenate protein. Touster et al. [40] suggest that both of these fractions represent plasma membrane. House et al. [41] have reached similar conclusions.

The ten-fold discrepancy between the predicted membrane yield calculations based on geometric data _versus_ marker enzyme data can be resolved by postulating one of the following possibilities.

1. A ten-fold increase in membrane surface area over the geometric calculation exists, and vesiculation causes most of the plasma membrane to appear in a fraction of smooth vesicles.

2. 5' -nucleotidase is localized in significant amounts in intracellular membranes as well as surface membranes.

A decision between possibilities 1 and 2 cannot be made at this time. Investigators should be cautioned that smooth membrane fractions enriched in 5'-nucleotidase may contain significant contributions from internal membranes.

C. Other Aids in Assessing Purity

The phase microscope is helpful in assessing purity when dealing with large membranes that have a distinctive morphology, such as liver [1] and brush border [38]. A general idea of the extent of small vesicular contamination owing to mitochondria or microsomes can be obtained. Although microsomal particles lie below the resolving power of the light microscope, they are visible as fine grains either from clumping or diffraction phenomena.

Acrylamide gel electrophoresis in sodium dodecyl sulfate is also a valuable tool in assessing purity. It has been shown that when the resolution of this technique is increased by the use of discontinuous buffers, each membraneous organelle has its own distinctive protein subunit pattern [4,5].

IV. GENERAL ISOLATION METHODS

There are three general methods for isolating plasma membranes. The choice of method is usually dictated by the cell type being used and the use to which the membranes will be put. The response of the plasma membrane to cell breakage is the crucial factor in the choice of method. Certain surface membranes vesiculate to small vesicles (2-0.1 microns), much the same as the tubular membranes of endoplasmic reticulum form microsomal vesicles on cell rupture. Morphologically, the smaller of these membranes are indistinguishable from smooth microsomes. Isolation procedures must rely on marker enzymes (see previous section) or some method of marking the outer surface of the cell [43]. The membrane vesicles are then separated by differences in density. Smooth membraneous vesicles are heterogeneous with respect to both size and density; however, discrete fractions can be isolated by density separations in sucrose [44]. The use of high-molecular-weight polymers, e.g., dextran, which do not osmotically shrink the vesicles, may improve the fractionation. This method was developed by Wallach and Kamet [45]. Plasma-membrane fractions isolated by this method will have as their major contaminant, cytoplasmic smooth vesicles similar in general composition to plasma membranes. Since their enzymatic composition also may be similar, it will be difficult to quantitate what fraction of the vesicles was of plasma-membrane origin. A variant of this method is to alter the homogenizing media so that the membrane vesiculates to large vesicles 5-10 microns in diameter. This is achieved by varying the calcium and magnesium concentration, added albumin concentration, temperature, and shear. These large vesicles can be identified by phase microscopy and separated by their higher sedimentation coefficient. Some smaller vesicles remain inside these ghosts. This technique has been successfully applied to fat cells by Rodbell [46].

Plasma membranes that vesiculate on cell rupture can be "toughened" by reacting the cells with agents that complex free sulfhydryl groups prior to homogenization. Cell rupture now releases the membrane as an intact sheet and the membranes can be identified morphologically. A minimal amount of sulfhydryl reagent is used to prevent clumps of cytoplasm from adhering to the membrane. Some enzymatic markers may be lost by this technique; however, the ease of isolating large sheets of identifiable membranes make this technique easier to apply than the small-vesicle techniques. Separation of membranous sheets from homogenate can be accomplished by a variety of methods, depending on the cell type used. This method was developed by Warren et al. [47].

A third technique is applicable when cell rupture in hypotonic media (<20 mM monovalent cation) releases intact sheets of plasma membranes. Under these conditions, nuclei and nuclear membranes are largely solubilized. The cell membranes are the largest fragments in the homogenate and can be concentrated by a low-speed centrifugation. Mitochondria and microsomal fragments are generally more dense and smaller and are preferentially eliminated by floating membranes in sucrose just above their isopycnic density and then performing a zonal centrifugation stabilized by a sucrose gradient. The smaller microsomal and mitochondrial contaminants are left near the application zone while the larger membranes are recovered from a dense cushion of sucrose at the bottom of the gradient. The advantage of this method is that it allows both morphological and biochemical evaluations of purity [1,12].

When a new membrane-isolation procedure is applied, the following method can be used to determine the response of the membrane to cell rupture. Homogenize until one-half the cells are ruptured. This will leave some cells with their membrane broken but their contents still clumped together. If the membrane remains as a sheet, it will be seen under phase microscopy on the periphery of the cell. By a tap of the cover slip, the membrane may be dislodged and seen to float away [1]. If the membrane vesiculates, sheets will not be seen. This procedure should be done in a cold room since membranes at room temperature stick to glass and lose identifiable morphology. Care must be taken not to confuse nuclear membranes for surface membranes. An enzymatic marker for nuclear membranes has been identified in certain cells [48].

It should be emphasized that although each general technique has been used on various cell types, the particular conditions used must be optimized for the particular cell. Plasma membranes and cytoplasmic organelles from different cell types not only vary in their size and exact density, but also in their susceptibility to divalent cations and sulfhydryl reagents. In addition, it should be noted that many tissues such as adipose tissue have basement membranes which surround large surface areas of cells and are in intimate contact with the adjacent cell membrane. In general the cells must first be freed from the basement membrane before homogenization, or on cell rupture the plasma membrane remains adherent to the basement membrane. This may necessitate making a suspension of cells by prior incubation with the mixtures of chelating agents, hyaluronidase, and collagenase [46].

The isolation of cell membranes from solid tumors deserves special mention. In our experience many tumors contain significant numbers of dead cells even though necrosis is not grossly apparent. On homogenization, the cytoplasmic particles of dead cells clump with membranes so that the membranes have a speckled appearance. Once clumped, these vesicles cannot be removed and high contamination rates may be unavoidable.

V. MOUSE LIVER PLASMA MEMBRANE ISOLATION

A. Comparison of Plasma Membranes Isolated from ob/ob Mice and Their Thin Litter Mates

The method used is presented in the appendix and is the third type of general method described in the previous section. The mouse strain used was C57BL/6J. Mice with the obese hyperglycemic syndrome carried a double dose of the recessive mutation

It was found that essentially the same method used for rat liver was applicable to mouse liver [49]. By using 10 g of tissue from each group, a 4-place rotor for the first sedimentation, and a 3-place rotor for the flotation and zonal centrifugation, the control group is centrifuged with the experimental group, thus minimizing procedural differences.

Examination of Table 1 shows that the yield of membrane protein, the enrichment of the membrane marker 5'-nucleotidase, and the contamination rates of the major liver organelles are comparable for the obese mouse and its thin litter mates. This is in spite of a gross difference in the starting material. In addition, we note that at each step in the purification process the yields of protein, plasma membrane, microsomes, and mitochondria are comparable for the two groups. This indicates that the fractionation procedure is not selecting differentially with respect to the thin and fatty livers, at least by these criteria. In addition, as seen in Fig. 1, the purified membranes of each group are indistinguishable by phase microscopy.

The problem of selection of various parts of the plasma membrane is more difficult to approach. The same specific activity of the membrane marker 5'-nucleotidase in both groups gives confidence that extensive selection is not taking place since Evans [37] has shown that shear-induced subfractions of liver-plasma membranes have different specific activities of this enzyme. The subfractions described by Evans [37] had dif-

TABLE 1

Comparison of Plasma Membrane Purification of ob/ob Mice and Their Thin Litter Mates. (All results are the average of 2 or 3 determinations.)

Purification fraction (step in Appendix) Part A	ob/ob			Thin		
	Total protein (mg)	5'-Nucleotidase (μmoles/mg/hr)	Relative specific activity	Total[a] protein	5'-Nucleotidase[b] (μmoles/mg/hr)	Relative specific activity
Dilute homogenate (2)	1450	0.30	1.0	1517	0.35	1.0
First sediment (3)	110	1.9	6.3	101	1.9	5.4
Floated particles (11)	21	4.3	14	24	5.0	14
Purified membranes (15)	5.3	6.8	23	6.9	6.8	19

[a] The total Lowry protein is expressed as yield in mg per 10 g wet weight of liver (albumin standard) [52].

[b] The 5'-nucleotidase activity was determined at 37°C as described [51] with the addition of a deproteinization step.

TABLE 1 cont'd

Purification fraction (step in Appendix)	ob/ob			Thin		
Part B	Succinate cytochrome C reductase[c] (μmoles/mg/min)	Relative specific activity	Percent mitochondrial protein in fraction	Succinate cytochrome C reductase[c] (μmoles/mg min)	Relative specific activity	Percent mitochondrial protein in fraction
Dilute homogenate (2)	0.027	1.0	15	0.035	1.0	15
First sediment (3)	0.043	1.6	24	0.061	1.7	26
Floated particles (11)	0.036	1.3	20	0.058	1.7	26
Purified membranes (15)	0.005	0.19	2.9	0.006	0.17	2.6

TABLE 1 cont'd

Purification fraction (step in Appendix)	ob/ob			Thin		
Part C	Glucose-6-phosphatase[d] (μmoles/mg/hr)	Relative specific activity	Percent microsomal protein in fraction	Glucose-6-phosphatase[d] (μmoles/mg/hr)	Relative specific activity	Percent microsomal protein in fraction
Dilute homogenate (2)	4.9	1.0	25	5.2	1.0	25
First sediment (3)	7.0	1.4	35	5.4	1.0	25
Floated particles (11)	0.84	0.17	4.3	1.1	0.21	5.3
Purified membranes (15)	0.42	0.09	2.3	0.54	0.10	2.5

ferent protein subunit compositions, as revealed by acrylamide gel electrophoresis in the presence of sodium dodecyl sulfate. The gel electrophoretic patterns for both groups of purified membranes are shown in Fig. 2 and are remarkably similar.

Another way to minimize the risk of differential plasma-membrane selection between two groups is to study another receptor or component which is most similar to the receptor in question. In the present case we were interested in assessing the role of surface-membrane insulin-receptor concentration in the insulin-resistant state characteristic of the hyperglycemic mouse [36]. We chose the glucagon receptor as a reference receptor. Although glucagon has opposite effects from those of insulin on the liver, it, like insulin, is a small polypeptide hormone which is secreted by the pancreas directly into the portal blood circulation. In Fig. 3 it can be seen that although the concentration of glucagon receptors per mg of membrane protein falls about 20% in the ob/ob membranes, the fall in insulin receptors is 75%.

B. Comments on the Method

The hypotonic media devoid of divalent cations allow for cell breakage to be achieved with a minimum of shear, thereby minimizing vesiculation of membranes which would be lost to the microsomal fraction. Homogenization should achieve at least

99% cell rupture; otherwise, cells will be floated and cytoplasmic contaminants will be carried through to the final step. Cell breakage is monitored by phase microscopy. Pestles of Dounce-type homogenizers often vary in diameter and should be standardized by timing the free descent of the pestle in the water-filled homogenizer. The dilution with media (step 2 of Appendix) aids in dispersion of clumped cytoplasm, dissolution of nuclei and nuclear membranes, and the subsequent sedimentation of membranes.

The centrifugal field (step 3 of Appendix) is chosen to maximize the sedimentation of the large membranes and minimize the sedimentation of mitochondria and clumps of endoplasmic reticulum. Angle-head rotors and plastic bottles cannot be used for this step, owing to inadequate packing of the pellet.

The choice of density for flotation (steps 6-8 of Appendix) is such as to maximize flotation of membranes and minimize flotation of mitochondria and microsomes. The density chosen is often quite critical because, although mitochondria, for example, have a higher average density than plasma membranes, their density distribution does overlap, and there are 1000 mitochondria for each plasma membrane. Flotation through the top layer of sucrose minimizes the flotation of microsomes and mitochondria. This step is an equilibrium flotation with respect to the large membranes, but the smaller particles have insufficient time to reach equilibrium. Reproducibility is ensured by the use of a refractometer. The density may need to be varied to suit species, nutrition, and other differences. Eighty percent of the membranes should float to form a skin at the liquid-air interface. If membranes do not lie evenly on the cushion, the linearity of the gradient should be checked.

APPENDIX

Mouse (C57BL/6J) liver-cell-plasma membranes were isolated using a modification of the author's original technique [1,49]. The method is described below in detail because attention to these details is necessary to ensure reproducibility. In particular, the sucrose concentrations are quite critical. These are given here in percent (w/w) at 20°C and are checked by an Abbe refractometer. All steps are carried out between 0° and 4°C. Medium refers to 0.001 M $NaHCO_3$. Three 10-g lots of liver can be processed together.

1. Decapitate enough mice to provide 10 g of liver. Trim livers free of the gall bladder (without rupture) and connective tissue, place in an iced beaker, and mince with scissors.

2. Place 10 g of minced liver in a large Dounce homogenizer (available from Blaessig Glass Company, Rochester, N. Y.). Add 25 ml medium. Homogenize at 4°C with 7 vigorous strokes of the loose pestle. Add 250 ml medium (4) and stir for 2 min, then filter first through 2 layers of cheesecloth, then through 4 layers.

3. Distribute the filtered homogenate equally between two 250-ml glass centrifuge bottles (do not prechill the bottles) and spin 1500 x g max for 10 min using a swinging bucket rotor. Use a speed of 2800 rpm for a Head #947 IEC B-20 centrifuge or Sorvall RC-2 using head #HS-4, or 2700 rpm for IEC PR-2 centrifuge using head #284. To accommodate HS-4 or 947 heads, the tops of 250 ml Kimax bottles (Kimble #14700) are sawed off leaving a cup which stands 11 cm high.

4. Carefully pour off supernatant and, while tube is upside down, insert absorbent paper into neck to remove excess supernatant. Now pour off pellets into a large Dounce homogenizer.

5. Homogenize pellets with three gentle strokes of the loose pestle.

6. Adjust 69% stock sucrose solution to 69 $\pm$ 0.1% using a refractometer. Place 11 ml into a 50-ml cylinder and cool in an ice bucket. Pour in the homogenate from step 5. Add H_2O (4°) to make 20 ml. Mix vigorously with rod until no Schlerin patterns are evident. Check the mixture in the refractometer. Adjust with H_2O or 69% sucrose until the homogenate sucrose concentration reads 44.0 $\pm$ 0.1%.

7. Pour 20 ml into an S-25.1 tube (Beckman Instruments, Inc.).

8. Carefully overlay this amount with 10 ml of 42.3 $\pm$ 0.1% sucrose (checked by the refractometer).

9. Balance all three tubes with ±0.05 g by adding 42.3% sucrose.

10. Load the tubes into a prechilled swinging bucket rotor (SW 25.1) and spin 25,000 rpm (90,000 x g max) for 2 hr, with the brake on. Handle the tubes carefully to preserve the density interface. Make certain that the caps are well greased to keep the tubes at 1 atm. (See step 13 now.)

11. Remove the float with a spatula. Add 8 ml medium. Spin, to pack the sediment, 25,000 x g max for 10 min.

12. Add 1 ml medium and resuspend the pellet by squirting once through a 4-in. No. 22 needle.

13. Into the bottom of a siliconized S-25.1 tube (Siliclad, Clay Adams, Parsippany, N. J., 07054), place a "cushion" of 4.1 ml of 50 ± 1% sucrose. Over each cushion form a linear sucrose gradient from 27-3%. For a Beckman gradient former, use the following solutions and settings: heavy 30%, light 0%, 50 ml syringes containing 20 ml, hold up -8, start 0, end 62. Run on slow.

14. Overlay the gradient with the resuspended homogenate (~1.3 ml) and spin 2000 rpm (550 x g max) for 1 hr with the brake off.

15. Using a syringe and long blunt No. 20 needle, remove the material at the cushion interface. Dilute with the medium to make 9.2 ml. Place 1.5 ml into 6 small conical sealable plastic centrifuge tubes (Eppendorf) and centrifuge at 15,000 x g

for 3 min. Discard the supernatant and freeze the pellet at -70°C. Examine the remainder of the diluted membranes by phase microscopy.

REFERENCES

1. D. M. Neville, Jr., J. Biophys. Biochem. Cytol., 8, 413 (1960).

2. E. L. Beneditti and P. Emmelot in The Membranes, A.J. Dalton and F. Haguenau, eds., Academic Press, New York, 1968, p. 33.

3. E. D. Korn, Ann. Rev. Biochem., 38, 263 (1969).

4. D. M. Neville, Jr. and H. Glossmann, J. Biol. Chem., 246, 6335 (1971).

5. H. Glossmann and D. M. Neville, Jr., J. Biol. Chem., 246, 6339 (1971).

6. J. N. Langley, J. Physiol., 33, 374 (1905).

7. P. Weiss, Yale J. Biol. Med., 19, 235 (1947).

8. O. Hechter, Cancer Res., 17, 512 (1957).

9. W. C. Stadie, N. Haugaard, J. B. Marsh, and A. G. Hills, Am. J. Med. Sci., 218, 265 (1949).

10. E. W. Sutherland, T. W. Rall, and T. Menon, J. Biol. Chem., 237, 1220 (1962).

11. J. P. Jost and H. V. Rickenberg, in Ann. Rev. Biochem. (E. E. Snell, P. D. Boyer, A. Meister, and R. L. Sinsheimer, eds.), Vol. 40, 1971, p. 741.

12. S. L. Pohl, L. Birnbaumer, and M. Rodbell, J. Biol. Chem., 246, 1849 (1971).

13. R. J. Lefkowitz, J. Roth, W. Pricer, and I. Pastan, Proc. Natl. Acad. Sci. U. S., 65, 745 (1970).

14. S. Y. Lin and T. L. Goodfriend, Amer. J. Physiol., 218, 1319 (1970).

15. M. Rodbell, H. M. J. Krans, S. L. Pohl, and L. Birnbaumer, J. Biol. Chem., 246, 1861 (1971).

16. P. D. R. House and M. J. Weidemann, Biochem. Biophys. Res. Commun., 41, 541 (1970).

17. P. Freychet, J. Roth, and D. M. Neville, Jr., Proc. Natl. Acad. Sci., U. S., 68, 1833 (1971).

18. P. Cuatrecasas, Proc. Natl. Acad. Sci., U. S., 68, 1264 (1971).

19. W. Stauffacher, L. Orci, D. P. Cameron, I. M. Burr, and A. E. Renold, Recent Progr. Hormone Res., 27, 41 (1971).

20. C. de Duve and J. Berthet, Intern. Rev. Cytol., 3, 225 (1954).

21. C. de Duve, B. C. Pressman, R. Gianetto, R. Wattiaux, and F. Appelmans, Biochem. J., 60, 604 (1955).

22. S. Wattiaux-DeConinck, M. J. Rutgeerts, and R. Wattiaux, Biochim. Biophys. Acta, 105, 446 (1965).

23. G. L. Sottocasa, B. Kylenstierna, L. Ernster, and A. Bergstrand, in Methods in Enzymology (R. W. Estabrook and M. E. Pullman, eds.), Vol. 10, Academic Press, New York, 1967, p. 448.

24. G. Hubscher and G. R. West, Nature, 205, 799 (1965).

25. H. Glossmann and D. M. Neville, Jr., FEBS Letters, 19, 340 (1972).

26. G. Weber and A. Cantero, Cancer Res., 15, 105 (1955).

27. D. Miller and R. K. Crane, Biochim. Biophys. Acta, 52, 293 (1961).

28. M. L. Clark, H. C. Lanz, and J. R. Senior, Biochim. Biophys. Acta, 183, 233 (1969).

29. D. Allan and M. J. Crumpton, Biochem. J., 120, 133 (1970).

30. D. M. Brunette and J. E. Till, J. Memb. Biol., 5, 215 (1971).

31. J. F. Perdue, R. Kletzien, K. Miller, G. Pridmore, and V. L. Wray, Biochim. Biophys. Acta, 249, 435 (1971).

32. P. H. Atkinson and D. F. Summers, J. Biol. Chem., 246, 5162 (1971).

33. A. M. Kidwai, M. A. Radcliffe, G. Duchon, and E. E. Daniel, Biochem. Biophys. Res. Commun., 45, 901 (1971).

34. I. G. Morgan, L. S. Wolfe, P. Mandel, and G. Gombos, Biochim. Biophys. Acta, 241, 737 (1971).

35. A. M. Kidwai, M. A. Radcliffe, and E. E. Daniel, Biochim. Biophys. Acta, 233, 538 (1971).

36. C. R. Kahn, D. M. Neville, Jr., and J. Roth, J. Biol. Chem., 248, 244 (1973).

37. W. H. Evans, Biochem. J., 116, 833 (1970).

38. R. Wilfong and D. M. Neville, Jr., J. Biol. Chem., 245, 6106 (1970).

39. H. Elias and J. C. Sherrick, Morphology of the Liver, Academic Press, New York and London, 1969.

40. O. Touster, N. N. Aronson, Jr., J. T. Dulaney, and H. Hendrickson, J. Cell Biol., 47, 604 (1970).

41. P. D. R. House, P. Poulis, and M. J. Weidemann, Eur. J. Biochem., 24, 429 (1972).

42. C. C. Widnell, J. Cell Biol., 52, 542 (1972).

43. M. S. Bretscher, J. Mol. Biol., 58, 775 (1971).

44. J. Rothschild, Biochem. Soc. Symp., 22, 4 (1963).

45. D. F. H. Wallach and V. B. Kamet, Proc. Natl. Acad. Sci., U. S., 52, 721 (1964).

46. M. Rodbell, J. Biol. Chem., 242, 5744 (1967).

47. L. Warren, M. C. Glick, and M. K. Nass, J. Cellular Comp. Physiol., 68, 269 (1966).

48. C. B. Kasper, J. Biol. Chem., 246, 577 (1971).

49. D. M. Neville, Jr., Biochim. Biophys. Acta, 154, 540 (1968).

50. H. D. Tisdale in Methods in Enzymology (R. W. Estabrook and M. E. Pullman, eds.), Vol. 10, Academic Press, New York, 1967, p. 213.

51. C. C. Widnell and J. C. Unkeless, Proc. Natl. Acad. Sci., U. S., 61, 1050 (1968).

52. O. H. Lowry, N. J. Rosebrough, A. L. Farr, and R. J. Randall, J. Biol. Chem., 193, 265 (1951).

53. D. M. Neville, Jr., J. Biol. Chem., 246, 6328 (1971).

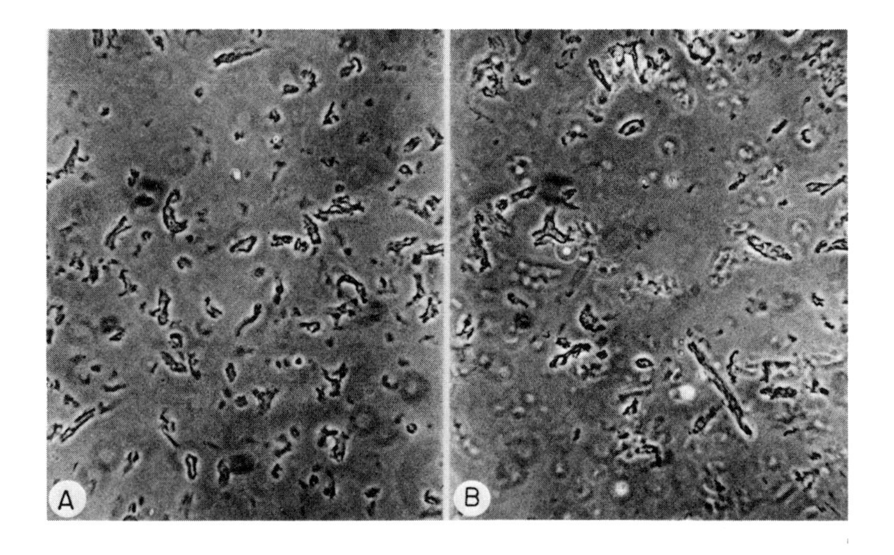

FIG. 1. Appearance of the purified membranes using phase microscopy. Membranes from the obese mouse are on the right (B), from the thin mouse on the left (A). Base magnification x 250.

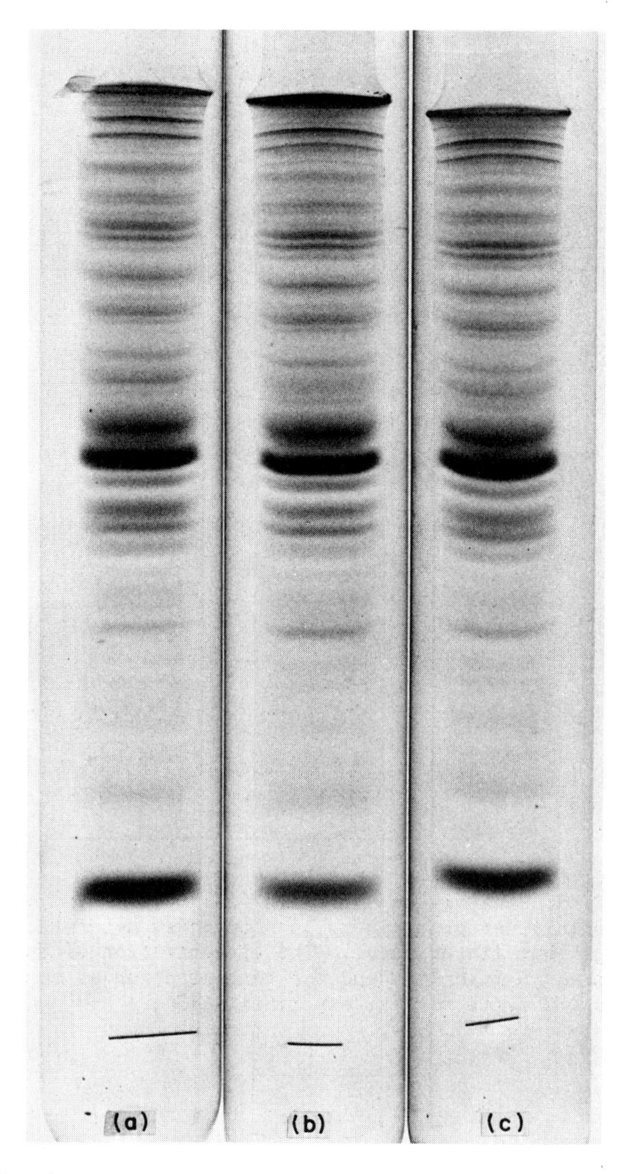

FIG. 2. Polyacrylamide gel electrophoresis in SDS. From left to right the gels are from membranes of the obese (A), the thin (B), and a mixture of both thin and obese (C). Each gel exhibits 40 resolvable bands. There are no bands observed in one membrane which are not observed in the others, and 2 of the 40 differ slightly in intensity. The gels contain 11% acrylamide and are run with a discontinuous buffer system at pH 9.5 as described [53].

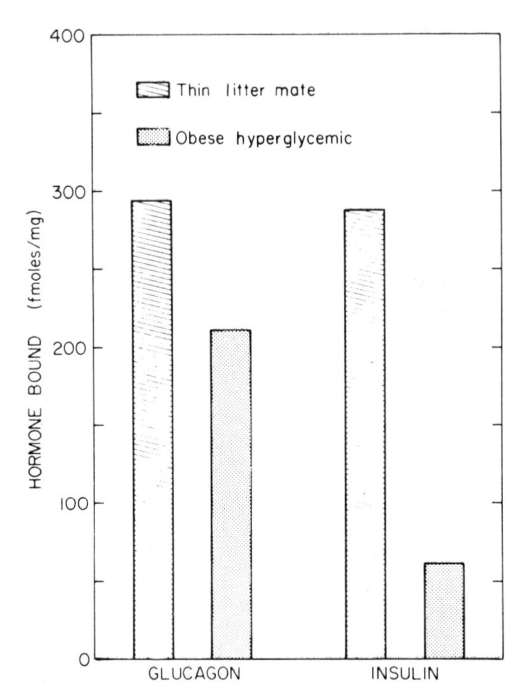

FIG. 3. Comparison of binding of ^{125}I-insulin and ^{125}I-glucagon to liver plasma membranes of obese hyperglycemic mice and their thin litter mates. The concentration of insulin and glucagon was 5×10^{-10} M and the concentration of membrane protein was 0.2 mg/ml for all experiments [36].

Chapter 4

ISOLATION OF LYSOSOMES

Gerald L. Rowin

Wallace Laboratories
Cranbury, New Jersey

I. INTRODUCTION

The lysosome as a distinct subcellular particle was demonstrated in 1955 by de Duve's group at the University of Louvain [1] and the name lysosome was given to this unique particle by de Duve (see Ref. [2]). The lysosome is a fragile membrane-bound particle containing a host of degradative enzymes (glycosidases, nucleases, proteases, etc.) that are capable of degrading a wide variety of biological molecules, bacteria, intracellular structures, and foreign substances. The lysosome has received a great deal of attention in the last fifteen years, most notably in the laboratories of de Duve (University of Louvain and Rockefeller University), Weissmann (New York University School of Medicine), and Tappel (University of California). Biochemical research on the lysosome has been directed mainly toward enzymological studies, the relationship of the lysosome to the pathological state, particularly arthritis, and purification of the lysosomal particle for characterization purposes.

Lysosomal fractions are invariably prepared by some type of centrifugation technique. Beaufay [3] has recently presented an excellent critical review of these methods with particular emphasis on the theoretical limitations of them. To date it has not been possible to prepare pure lysosomes as the preparations are invariably contaminated to different degrees with mitochondria, peroxisomes, microsomes, or combinations of these subcellular particles. It is, therefore, best to speak of a lysosomal fraction even though it may be highly purified or contain only a small percentage of lysosomes. Lysosomal fractions are useful starting points for purification of lysosomal enzymes [4-11], study of the *in vitro* interaction of drugs with lysosomes [12-17], and further purification of lysosomes [18-20].

The choice of isolation method should be based on the ultimate use intended for the lysosomal fraction. This chapter will attempt to help the researcher gain a foothold in lysosomal methodology and indicate to him the pathways available once his techniques are in hand.

II. HANDLING OF TISSUES

The most studied tissue used for fractionation of subcellular particles has been the rat liver. General comments concerning the handling of liver tissue apply equally well to other tissues. Because of the fragility of the lysosomal particle to hypotonic media, freezing and thawing, and increased temperature, special precautions must be taken to avoid these extremes after collection of the tissue. Animals should be healthy and fasted at least 12 hours prior to sacrifice. They should be killed by decapitation, cervical dislocation, or a blow on the head rather than with ether or chloroform inhalation.

The tissue should be quickly and neatly removed and immediately placed in ice-cold medium. The medium of choice for most tissues is 0.25 M sucrose. The 0.25 M sucrose may contain such additives as 0.001 M EDTA, pH 7.0-7.2, or 0.01 M Tris-HCl, pH 7.1. The transport of abattoir materials to the laboratory requires special attention, e.g., Barrett [5] transported minced rabbit liver in a thermos containing a 0.25 M sucrose slush thus diminishing the possibility of lysosomal labilization prior to workup. Working temperatures should be kept as near 0°C as possible during all phases of the experimentation.

III. HOMOGENIZATION OF THE TISSUE

Homogenization of liver and other noncartilaginous tissues is best performed in a precision-bore tube fitted with either a Teflon or Kel-F pestle. The tougher tissues may require the all-glass Potter-Elvehjem type tissue grinder. This latter type of grinder has the disadvantage of shedding glass particles into the homogenate, and also the clearance between the pestle and tube changes with use. The Dounce type homogenizer is useful not only for grinding of tissues but also for resuspension of precipitates. It is available with two sizes of pestles. All these grinders are available from the Kontes Glass Co., Vineland, New Jersey. A 1/2-inch drill mounted on a drill-press stand clamped to the bench top serves as an ideal source for driving the Teflon or Kel-F pestle. It has sufficient torque to drive the pestle through the tissue without slowing down. It can also be adapted for use with the Potter-Elvehjem grinder. To keep the tube cold during the homogenization, it is immersed in a plastic beaker, or wash bottle with the top cut off, that is filled with an ice slush. It is essential that the homogenization be performed as methodically as possible, as subsequent adjustments of the total fractionation procedure depend upon a uniform homogenization program.

IV. ENZYME ASSAYS

A comprehensive list and review of the properties of lysosomal enzymes, including references for their determination, is given by Barrett [21]. The success of the fractionation procedure is judged by determination of marker enzymes, i.e., enzymes known to be associated specifically with a single subcellular fraction. Use of these marker enzymes allows quantitative determination of the contamination of one subcellular fraction by another. The accepted marker enzymes for the various subcellular particles and references for determination of them are: mitochondria--cytochrome oxidase [22]; lysosomes--acid phosphatase [23]; peroxisomes--catalase [24]; microsomes--glucose-6-phosphatase [2].

The enzyme assays involve various types of spectrophotometric determinations and any type of good instrument will suffice for this purpose. It has been found convenient to prepare a series of matched tubes (12 x 75 mm) from stock for use with colorimetric procedures in the visible light range. These tubes should be kept in rubber-coated or plastic racks and washed only with cotton-tipped wooden sticks to prevent scratching them. These tubes are acceptable in the Coleman Jr. Spectrophotometer fitted with the proper adapter. Many of the enzyme assays involve the determination of liberated _p_-nitrophenol. These assays are particularly convenient and a good source for the appropriate _p_-nitrophenyl substrates is the Pierce Chemical Co., Rockford, Ill. A few general comments about the enzyme assays follow.

1. All determinations should be performed in duplicate.

2. Incubations are usually made in stoppered tubes at 37°C in a thermostatically controlled water bath. If the incubation period exceeds 2 hours, the reactions should be done in an air incubator with a drop of toluene as preservative.

3. Substrate blanks, enzyme blanks, and the standards should be incubated simultaneously with the reaction mixtures.

4. The pH of stock buffer solutions should be adjusted so that they give the proper pH when diluted to assay concentration.

5. When the volume of liquid to be measured is less than 0.5 ml, micropipettes should be used.

6. One unit of enzyme activity should be described as that amount (gram of tissue or ml of preparation) that will cleave 1 micromole of substrate per minute.

7. It should be established that the enzymic reaction is linear during the assay period, linear with enzyme concentration, and does not exceed 5-10% cleavage of the substrate.

8. If the presence of sucrose interferes with the enzyme assay, the particle may be pelleted by centrifugation and resuspended in an appropriate medium.

9. When higher concentrations of tissue are used to give demonstrable enzymic activity, turbidity occasionally develops in the final solution. This frequently can be removed by the addition of a drop of Triton X-100, centrifugation, or extraction with pentanol-chloroform [25].

V. DIFFERENTIAL CENTRIFUGATION OF THE HOMOGENATE

The fractionation procedure of most general applicability is the method of de Duve et al. [2]. With this method the lysosomal fraction is purified about fivefold over the homogenate and contains approximately 40% of the lysosomal enzymes of the homogenate. The procedure not only yields an enriched lysosomal fraction but also allows determination of the subcellular fraction in which the material under investigation (enzyme, protein, lipid, etc.) resides. From the homogenate, four particulate fractions are collected. They are: nuclear = N; heavy mitochondrial = HM; light mitochondrial (lysosomal) = LM; and microsomal = M. In addition, samples of the cytoplasmic extract = CE and soluble = S fractions are reserved to completely monitor the fractionation.

Prior to the fractionation it is necessary to tare several 50-ml centrifuge tubes, as subsequent dilutions are made on a weight basis. The polycarbonate tubes assure easy visibility of the contents. When processing smaller amounts of tissue, the high-speed Corex tubes (Corning Glass Co., Corning, New York) are useful.

The N, HM, and LM fractions are prepared in the SS-34 rotor of the Sorvall RC2-B refrigerated centrifuge (set at 0°C). The microsomal and soluble fractions are obtained in the Spinco Model L or L2-65B fitted with the Type 40 or Type 65 rotors, respectively. The centrifugation conditions are given as rpm per time interval at the plateau speed. The _g_ values refer to the bottom of the tube for the SS-34 rotor and to the middle of the tube for the Spinco rotors. The centrifugal field generated during acceleration and deceleration of the rotors is not considered; however, for discussions of these factors, see Refs. [26-28].

After the tissue is removed, it should be placed in a tared beaker of ice-cold medium (medium here refers to 0.25 _M_ sucrose with or without additives as previously described) and the weight of the tissue determined. When processing tissue-culture cells, it is convenient to measure the packed-cell volume. With leucocytes, a cell count can be made and the experiment can be based on the number of cells per ml. The tissue is then minced with scissors and the medium decanted. A further rinse of the tissue will serve to remove blood components. The tissue is then blotted with a Kim-Wipe, transferred to the homogenizer tube, and an amount of medium equal to three times the weight of the tissue is added. The tissue is homogenized with a steady upward thrust until the pestle has reached bottom. The pestle is then slowly brought back through the medium. The homogenate is centrifuged at 2700 rpm (1000 x _g_) for 10 min in a tared tube. The supernatant fluid is carefully removed with a large pipette fitted with a Propipette. The crude nuclear precipitate is homogenized and cntrifuged as above two additional times. The nuclear pellet (N) is kept on ice and the combined supernatant fluids constitute the CE. An accurately measured aliquot (10-20%) of the CE is kept for later analysis and the remainder is centrifuged at 4900 rpm (3300 x _g_) for 10 min in a tared tube. The supernatant fluid is very carefully removed by pipette (consistency in removal of the supernatant fluid at this point is essential) and the pellet is washed twice by gentle resuspension (by glass rod or pipetting) in 10 ml of medium followed by centrifugation as before. The pellet (HM) is iced and the combined supernatant fluids are centrifuged at 14,300 rpm (25,000 x *g*) for 10 min. The lysosomal pellet (LM) is washed in the same manner as the HM pellet. The combined supernatant fluids are centrifuged at 40,000 rpm (100,000 x *g*) for 30 min in the Type 40 rotor or 65,000 rpm (about 300,000 x *g*) for 15 min in the Type 65 rotor. The soluble fraction (S) is easily decanted from the firmly packed microsomal pellet (M). To avoid further dilution of the S fraction, the microsomal pellet is not washed. For bookkeeping purposes it must be determined what fraction of the LM supernatant fluid was processed in the ultracentrifuge because of the necessity to completely fill the tubes, with the result that there is always some fluid left unprocessed.

After completion of the fractionation, the N, HM, LM, and M fractions are resuspended by gentle homogenization in medium using the Dounce type homogenizer with the loose-fitting pestle. These suspensions constitute the stock enzymes to be further diluted for assay. The dilution factors for the various fractions are expressed with regard to the weight of original tissue represented in the fraction, e.g., a 1:4 dilution of the HM fraction represents a suspension whose volume (in ml) is four times the weight of the original tissue. The dilutions are made by weight in the tared tubes, assuming a density of 1.05 g per ml for these suspensions. It is convenient to dilute all the particulate fractions 1:4, i.e., 0.25 g tissue per ml. It must be remembered that sampling the CE has reduced the amount of original tissue carried through the fractionation and the inability to utilize all of the LM supernatant fluid in the ultracentrifuge has the same effect on the M fraction. A simple example will be helpful. If a 7-g liver is processed as above and 10% of the CE is set aside, then 0.90 x 7.0 = 6.3 g of liver is represented in the HM and LM fractions. Processing 50% of the LM supernatant fluid in the ultracentrifuge means that 0.50 x 6.3 = 3.15 g of liver is represented in the M fraction. The concentrations of tissue in the CE and S fractions are simply the weight of liver represented in the fraction divided by the volume.

For determination of the enzymic activity of the above fractions, they should be further diluted prior to assay. As the lysosomal enzymes do not exhibit their full activity in fresh preparations, particularly in sucrose solutions [29-31], they should be further diluted in water or 0.1% Triton X-100 (Rohm and Haas Co., Philadelphia, Pa.), frozen and thawed several times or incubated at slightly acid pH for a short period of time. Alternatively, Triton X-100 can be present in the assay mixture at a final concentration of 0.1-0.2%. It may be added with the substrate or buffer or all three may be combined for addition to the assay. The appropriate blanks will indicate what mixtures are compatible. When extended incubation periods are required to display measurable enzymic activity, the lysosomal particle will usually rupture and total activity will be measured, owing to the presence of the acidic buffer and hypotonic conditions of the reaction. These conditions assure disruption of the lysosomal membrane, thus permitting availability of the enzyme for reaction with substrate. The enzyme may be solubilized by these procedures or remain bound to the membrane in an active form. Rat liver lysosomal β-acetylglucosaminidase [25] and β-glucosidase [32] have been shown to remain firmly bound to membrane after attempts to solubilize them.

The further dilutions given below will serve as a guide for initial experiments and may be altered as requirements indicate. They are: N (1:4), 1 to 80; CE, 1 to 200; HM (1:4), 1 to 100; LM (1:4), 1 to 200; M (1:4), 1 to 100; and S may be left undiluted or diluted 1 to 10. In tissues containing very low levels of enzyme, the dilution of the fractions should be substantially reduced. The quantitative analysis of the fractionation requires determination of the units of enzymic activity present in each fraction and expression of these values as a percentage of the sum of the N and CE (representing the whole homogenate).

Analysis of the results of the fractionation will indicate the changes to be made to optimize the conditions. For example, if the N fraction contains too high a percentage of lysosomal enzyme, it would indicate incomplete homogenization; and conversely, if the S fraction contains a high percentage of lysosomal enzyme, it would indicate that the particles have been ruptured by too vigorous a homogenization or that some endogenous factor is causing premature labilization of the lysosomes. Frequently the HM pellet contains more lysosomal enzyme than is necessary, but this can be reduced somewhat simply by sedimentation of the HM at slightly lower speed and use of greater care in the removal of the supernatant fluid.

A few examples of tissues other than rat liver from which an enriched lysosomal fraction has been prepared by the above procedure are as follows: rat kidney [33], rat brain [34], pig liver [25], guinea pig liver [35], human liver [36], human placenta [37], bone [38], Ehrlichs ascites tumor cells [39], and Walker 256 carcinoma [40].

The lysosomes of the rabbit polymorphonuclear leucocyte can be isolated by differential centrifugation [41]. The method requires the selection of rabbits able to produce PMN leucocytes (peritoneal) after priming with glycogen. Once selected, the rabbits produce these cells for extended periods of time. Preparation of cell homogenates is simplified by the fact that vigorous pipetting of the cells in 0.34 M sucrose causes them to lyse while at the same time provides an osmotically suitable medium for the lysosomes. The lysosomal fraction represents about the same yield and purification factor as the LM fraction of de Duve et al. [2]. However, the method has not been successful with human PMN leucocytes [42] or guinea pig PMN leucocytes [43].

Differential centrifugation has been used by Ragab et al. [44] to prepare a rat liver lysosomal fraction whose increase in specific activity (based on acid phosphatase) was about 24-fold. The yield, however, was very low. The procedure involves a series of centrifugations of the liver homogenate and subsequent fractions through layers of increasing concentrations of sucrose, using the Sorvall centrifuge fitted with the GSA rotor. Although a large amount of tissue can be processed at a single time (250-360 g of liver), the method is tedious and variable and the low yield is discouraging.

VI. TRITON-FILLED LYSOSOMES

In studying the effects of Triton WR-1339 on the liver lysosomes of rats previously injected intravenously with this detergent, Wattiaux et al. [18] noticed that the equilibrium density of the lysosomal fraction had been drastically altered, when compared to identical experiments on rats treated with saline. The lysosomes of the detergent-treated rats were larger and less dense than those of the untreated animals owing to the uptake of the detergent by the liver lysosomes. The increase in size of the particle with concomitant decrease in density did not affect the location of the lysosomes in the fractionation scheme of de Duve et al. [2]. These factors did, however, alter the position of the lysosomes in a linear sucrose gradient.

Trouet [19] extended the above observations to a method whereby relatively large quantities of highly purified liver lysosomes can be easily prepared from rats treated with Triton WR-1339. A suspension of the HM and LM fractions in sucrose (density 1.21) is overlaid with three sucrose solutions of decreasing density (1.155, 1.14, and 1.06) and centrifuged for 2 hr at 25,000 rpm. The lysosomes float through the discontinuous gradient and appear at the interface between the upper two layers. The lysosomal fraction is only minimally contaminated by mitochondria, peroxisomes, or microsomes. Unfortunately, very little experimental detail was available in the Trouet report.

A modification of the Trouet procedure was reported by Leighton et al. [20]. Rats are injected intraperitoneally with Triton WR-1339 (Ruger Chemical Co., Irvington-on-Hudson, New York), 0.85 g/kg, 3-1/2 days before killing. The combined HM and LM fractions are then prepared according to de Duve et al. [2] and suspended in 45.0% (w/w) sucrose (density 1.21) to give a suspension containing 1 g of liver per ml. Using the SW 25.2 rotor of the Spinco Model L2-HV, 25 ml of the particle suspension is layered under 20 ml of 34.5% (w/w) sucrose (density 1.155) and 10 ml of 14.3% (w/w) sucrose (density 1.06). After centrifugation at 25,000 rpm for 2 hr, the lysosomes float to the interface between the upper two layers. The tube is unloaded from the top with a special device by allowing 60.7% (w/w) sucrose to push out the less-dense contents. (A commercially available device such as this is produced by Beckman Instruments, Inc., Spinco

Div., Palo Alto, Cal.) The lysosomal fraction prepared by this procedure is purified about 40-fold over the homogenate and is relatively free of contaminating particles as judged by the use of marker enzymes.

Franson et al. [45] have reported the appearance of two distinct populations of lysosomes in liver homogenates of rats treated with Triton WR-1339. They attributed this to the differential uptake of the detergent by some of the lysosomes. Horvat et al. [46] prepared a lysosomal fraction from Ehrlich ascites tumor cells propagated in mice. The tumor lysosomes accumulated Triton WR-1339 when the mice were injected intraperitoneally with the detergent.

Lysosomal fractions prepared from animals treated with Triton WR-1339 still contain the detergent and are more fragile than lysosomal fractions prepared from untreated animals. Because of this their utility in studies involving the integrity of the lysosomal membrane would be compromised. The excellent separation of Triton-filled lysosomes from the other subcellular particles makes this a powerful tool for definitive localization of an enzyme in the lysosome.

VII. ZONAL ROTOR TECHNIQUES

Zonal rotor techniques for the isolation of lysosomal fractions will be mentioned only briefly here, as they are more easily acquired by personal instruction or by attendance at the courses offered by the Beckman Instrument Co.

Rahman et al. [47] fractionated rat liver homogenates in an A-XII rotor filled with a 10-43% linear sucrose gradient. Based on enzymic determinations of fractions collected from the gradient, they postulated that two different lysosomal populations had been fractionated. Brown [48] was able to separate rat liver mitochondria, peroxisomes, and lysosomes into distinct though overlapping peaks in a 0-15% Ficoll gradient containing 0.25 M sucrose throughout. A B-XXIII rotor was used for this separation. Schuel et al. [49] were able to obtain excellent separation of rat liver lysosomes from mitochondria using the A-XII rotor with a sucrose gradient which was partially discontinuous and partially linear. Baggiolini et al. [50] fractionated homogenates of rabbit heterophil leucocytes using the B-XIV rotor with a discontinuous sucrose gradient (0.45-0.80 M). Two lysosomal fractions were separated. Canonico and Bird [51] reported on the fractionation of skeletal-muscle homogenates in the Spinco B-XIV rotor utilizing a 30-50% linear sucrose gradient. They concluded that the two lysosomal fractions which were separated arose from different tissues.

VIII. CONTINUOUS ELECTROPHORESIS

A novel and little-studied method for the purification of lysosomes has been reported by Stahn et al. [52]. By using carrier-free continuous electrophoresis, they described the preparation of a lysosomal fraction containing acid phosphatase with a 40-fold purification over the homogenate; for β-glucuronidase, 65-fold; and for arylsulfatase, 240-fold. They were also able to demonstrate heterogeneity in the lysosomal population based on various ratios of enzyme content in different fractions. Rat liver was the tissue used in this work.

REFERENCES

1. F. Appelmans, R. Wattiaux, and C. de Duve, Biochem. J., 59 438 (1955).
2. C. de Duve, B. C. Pressman, R. Gianetto, R. Wattiaux, and F. Appelmans, Biochem. J., 60, 604 (1955).
3. H. Beaufay in Lysosomes in Biology and Pathology (J. T. Dingle and Honor B. Fell, eds.), Part 2, American Elsevier, New York, 1969, pp. 515-546.
4. S. Mahadevan and A. L. Tappel, J. Biol. Chem., 242, 4568 (1967).
5. A. J. Barrett, Biochem. J., 104, 601 (1967).
6. R. Brightwell and A. L. Tappel, Arch. Biochem. Biophys., 124, 333 (1968).
7. C. Arsenis and O. Touster, J. Biol. Chem., 243, 5702 (1968).

8. C. Beck, S. Mahadevan, R. Brightwell, C. J. Dillard, and A. L. Tappel, Arch. Biochem. Biophys., 128, 369 (1968).

9. P. L. Jeffrey, D. H. Brown, and B. I. Brown, Biochemistry, 9, 1403 (1970).

10. G. E. R. Hook, K. S. Dodgson, and F. A. Rose, Biochem. J., 119, 28P (1970).

11. J. T. Dulaney and O. Touster, J. Biol. Chem., 247, 1424 (1972).

12. C. de Duve in Subcellular Particles (T. Hayashi, ed.), Ronald Press, New York, 1959, pp. 128-159.

13. J. T. Dingle, Biochem. J., 79, 509 (1961).

14. C. de Duve, R. Wattiaux, and M. Wibo, Biochem. Pharmacol., 9, 97 (1962).

15. G. Weissmann, Biochem. Pharmacol., 14, 525 (1965).

16. A. Allison, Adv. Chemotherapy, 3, 253 (1968).

17. G. Weissmann in Lysosomes in Biology and Pathology (J. T. Dingle and Honor B. Fell, eds.), Part 2, American Elsevier, New York, 1969, pp. 276-295.

18. R. Wattiaux, M. Wibo, and P. B. Baudhuin in Ciba Foundation Symposium on Lysosomes (A. V. S. de Reuck and M. P. Cameron, eds.), Churchill, London, 1963, pp. 176-200.

19. A. Trouet, Arch. Intern. Physiol. Biochim., 72, 698 (1964).

20. F. Leighton, B. Poole, H. Beaufay, P. Baudhuin, J. W. Coffey, S. Fowler, and C. de Duve, J. Cell. Biol., 37, 482 (1968).

21. A. J. Barrett in Lysosomes in Biology and Pathology (J. T. Dingle and Honor B. Fell, eds.), Part 2, American Elsevier, New York, 1969, pp. 245-312.

22. C. J. Cooperstein and A. Lazarow, J. Biol. Chem., 189, 665 (1951).

23. R. Gianetto and C. de Duve, Biochem. J., 59, 433 (1955).

24. P. Baudhuin, H. Beaufay, Y. Rahman-Li, O. Z. Sellinger, R. Wattiaux, P. Jacques, and C. de Duve, Biochem. J., 92, 179 (1964).

25. B. Weissmann, G. Rowin, J. Marshall, and D. Friederici, Biochemistry, 7, 207 (1967).

26. C. de Duve and J. Berthet, Nature, 172, 1142 (1953).

27. C. de Duve and J. Berthet, Intern. Rev. Cytol., 3, 225 (1954).

28. C. de Duve, J. Berthet, and H. Beaufay, Progr. Biophys. Chem., 9, 325 (1959).

29. C. de Duve, J. Berthet, L. Berthet, and F. Appelmans, Nature, 167, 389 (1951).

30. J. Berthet and C. de Duve, Biochem. J., 50, 174 (1951).

31. J. Berthet, L. Berthet, F. Appelmans, and C. de Duve, Biochem. J., 50, 182 (1951).

32. C. Beck and A. L. Tappel, Biochim. Biophys. Acta., 151, 159 (1968).

33. S. Wattiaux-de Coninck, M. J. Rutgeerts, and R. Wattiaux, Biochem. Biophys. Acta, 105, 446 (1965).

34. W. N. Aldridge and M. K. Johnson, Biochem. J., 73, 270 (1959).

35. V. Patel and A. L. Tappel, Biochim. Biophys. Acta, 208, 163 (1970).

36. T. Schersten, Biochem. Biophys. Acta, 141, 144 (1967).

37. S. F. Contractor, Nature, 223, 1274 (1969).

38. G. Vaes and P. Jacques, Biochem. J., 97, 380 (1965).

39. A. Horvat and O. Touster, Biochim. Biophys. Acta, 148, 725 (1967).

40. Y. T. Li, S. C. Li, and M. R. Shetlar, Cancer Res., 25, 1225 (1965).

41. Z. A. Cohn and J. G. Hirsch, J. Exptl. Med., 112, 983 (1960).

42. R. Hirschhorn and G. Weissmann, Proc. Soc. Exptl. Biol. Med., 119, 36 (1965).

43. G. L. Rowin, unpublished results.

44. H. Ragab, C. Beck, C. Dillard, and A. L. Tappel, Biochim. Biophys. Acta, 148, 501 (1967).

45. R. Franson, M. Waite, and M. LaVia, Biochemistry, 10, 1942 (1971).

46. A. Horvat, J. Baxandall, and O. Touster, J. Cell. Biol., 42, 469 (1969).

47. Y. E. Rahman, J. F. Howe, S. L. Nance, and J. F. Thomson, Biochim. Biophys. Acta, 146, 484 (1967).

48. D. H. Brown, Biochim. Biophys. Acta, 162, 152 (1968).

49. H. Schuel, R. Schuel, and N. J. Unakar, Anal. Biochem., 25, 146 (1968).

50. M. Baggiolini, J. G. Hirsch, and C. de Duve, J. Cell. Biol., 40, 529 (1969).

51. P. G. Canonico and J. W. C. Bird, J. Cell. Biol., 45, 321, (1970).

52. R. Stahn, K. P. Maier, and K. Hannig, J. Cell. Biol., 46, 576 (1970).

Chapter 5

ISOLATION OF THE GOLGI APPARATUS

William P. Cunningham

Department of Genetics and Cell Biology
University of Minnesota
St. Paul, Minnesota

I. INTRODUCTION

The Golgi apparatus is the primary site of assembly and packaging for a wide variety of secretory products (for a review see Refs. [1-4]) in nearly all eukaryotic cells. These range from plant-cell-wall precursors to liver-cell very-low-density lipoproteins, to lysosomelike particles such as the sperm acrosome, pancreatic zymogen granules, and leucocyte autolytic granules. There is growing evidence, both from direct biochemical studies of isolated organelles [5-7] and from cytochemical studies [8-10], that one of the main functions of the Golgi apparatus is the attachment of carbohydrate moieties to mucopolysaccharide or glycoprotein secretory products. In addition, since secretion usually involves the production of membranous vesicles that are transported to other areas of the cell, there must be a flow of membrane material from the site of synthesis (probably the endoplasmic reticulum) through the Golgi apparatus and into other membrane systems such as the plasma membrane. During this process, transformations in both membrane structure and composition probably occur [11] that aid in directing the secretory product to the proper area in the cell and make possible subsequent fusion with the proper membrane systems. Another interesting property of the secretory vesicle membranes is their ability to withstand the lytic action of their contents, which are frequently lysosomelike.

The purpose in attempting to extract the Golgi apparatus from the cell is to study these essential processes in vitro where they are accessible to direct biochemical characterization and experimental manipulation. The information gained in such studies will also be valuable for comparisons of Golgi apparatus structure and composition with those of other cellular organelles.

In this chapter the terminology suggested by Mollenhauer and Morre [3] will be used. The Golgi apparatus (either singular or plural) is defined as a complex of flattened membranous sacs (cisternae) arranged in parallel arrays and surrounded by a reticulate network of tubules and vesicles that connect with the periphery of the cisternae. Usually several of these cisternal stacks (dictyosomes) are connected by means of the peripheral tubules in a large hemispherical complex (see Fig. 1). Often there is a morphological and functional polarity across the Golgi apparatus. The cisternae on the convex side of the organelle have characteristics similar to those of the endoplasmic reticulum (ER) and tend to be highly fenestrate and tubular [9,12,15]. They are referred to as the immature or entrance face of the complex. The concave side of the complex has cisternae that appear to give rise to secretory vesicles that often have specialized surfaces (smooth, fuzzy, spiney coated), and is referred to as the mature or secretory or exit face.

Since the Golgi apparatus is usually a large, loosely organized complex that is relatively unstable when the cell is damaged, it is often one of the most difficult of the cellular organelles to extract in reasonable quantity, purity, and state of preservation. The presence of digestive enzymes in secretory vesicles associated with the Golgi complex may be one of the main impediments to the successful extraction of these organelles from many cell types. In recent years, however, isolation techniques have been devised for the extraction of Golgi apparatus from a number of cell and tissue types, and it now appears possible to begin the elucidation of the activities of this interesting and important system.

II. ISOLATION METHODS: TISSUE CHOICE

There are two main factors to be considered in choosing a tissue for organelle fractionation. One is the intrinsic interest of the organelles themselves. For example, the epididymis Golgi apparatus is large and complex and may be involved in the important process of sperm capacitance; the pancreas Golgi apparatus has an important role in zymogen granule formation and has been studied extensively by cytochemical and histological techniques. The second factor that must be considered, however, is the accessibility of the tissue. Pancreas and epididymis both yield only small amounts of material for the inevitably low efficiency of the fractionation process. In addition, epididymis has a high lytic activity and is very difficult to homogenize because of its high connective-tissue content. By the time the cells are broken, the organelles are also destroyed. Liver and testis have yielded good Golgi apparatus preparations because they are soft tissues, easy to homogenize, and readily available in the laboratory in a fresh state and in relatively large quantities. For an interesting discussion of the development of isolation techniques and difficulties involved, see DeDuve [16].

A. Epididymis

The first reported successful isolation of Golgi apparatus was from epididymis [17,18]. Homogenization was carried out in a loose-fitting Potter-Elvjhem homogenizer in a medium containing 0.25 M sucrose and 0.35 N NaCl. The resolution of structures in these studies is not good, but that may be owing more to electron microscopic preparative techniques available at the time the studies were done than to the isolation techniques.

The epididymis Golgi apparatus, because of its large size and complex organization, would certainly be an interesting system to study. I have had very little success in numerous attempts to isolate epididymis Golgi apparatus. These organelles seem to fragment very readily into small vesicles, perhaps because of a high proteolytic enzyme activity in this tissue. It is interesting that the epididymis Golgi apparatus was reported by Kuff and Dalton [18] to collect at the 1.12-1.13 density interface in a discontinuous sucrose gradient. This is very similar to the density at which testis Golgi apparatus equilibrates, as will be discussed later.

B. Onion Stem

A procedure has been developed for isolation of plant Golgi apparatus using a low concentration of glutaraldehyde in the homogenizing medium in order to stabilize the cisternal membranes [19]. This procedure yields very good structural preservation and organelle purification. Of course, most enzyme activities are destroyed but the organelles are useful for chemical characterization. This technique has also been used successfully for studies involving *in vivo* incorporation of radioactively labeled plant-cell-wall precursors [20]. This approach is useful with tissues where little is known about the Golgi apparatus or how it will behave in isolation.

Procedure (from Morre et al. [19]; see also Ref. [21]): Green onions (*Allium cepa*) are purchased locally and stored at 4°C. A cone of stem tissue approximately 1 cm diameter and 1 cm high is excised from the stem region in the center of the bulb. Approximately 5 g of tissue is obtained from 30-50 onions. Homogenization is accomplished either in a Polytron (Brinkman Instruments) or by chopping with razor blades. The latter is facilitated by the motor-driven device described by Morre [21].

The homogenizing solution contains 0.5 M sucrose, 0.01 M sodium phosphate buffer, pH 6.8, 1% dextran (w/v) of 200,000 mol wt and 0.05 M glutaraldehyde. About 10-20 ml of solution is used for 5 g of tissue. The homogenate is filtered through Miracloth (Chicopee Mills, New York) or cheesecloth to remove cell walls and debris and then centrifuged at 7000 x g for 20 min. The supernatant is centrifuged at 10,000 x g for 30 min onto a 1.6-M sucrose cushion. The supernatant solution is removed and the layers of 1.0 ml each of 1.5 M, 1.25 M, and 0.5 M sucrose are added. The tubes are centrifuged for 3 hr at 100,000 x g (average). The dictyosomes collect at the 0.5/1.25 M sucrose interface and are removed with a Pasteur pipette. The sucrose is diluted with an equal amount of distilled water and the dictyosomes are sedimented at 35,000 x g for 30 min. The morphology of such a fraction has been described by Cunningham et al. [13].

Discussion: This procedure may be used without glutaraldehyde stabilization by adjustment of the centrifugation speeds [21], but a loss in structural preservation and purity results. This procedure has been applied also be other plant tissues with varied degrees of success (see Ref. [21]).

C. Rat Liver (from Morre et al. [22])

The preceding technique not only yields good enzyme purification and fair morphological preservation, but also has the advantage of using tissue easy to obtain and maintain under controlled laboratory conditions.

Procedure: Adult (200-300 g) male or female rats (Holtzman Co., Madison, Wis.), fed a standard diet, are either decapitated or are anesthetized by intraperitoneal injection of 0.5 to 1 ml pentobarbital (Nembutal) solution (20 mg/ml). The livers (approx. 10 g each) are drained of blood (by blocking the portal vein and hepatic artery and severing the vena cava before excising the liver in the case of anesthetized animals) and rinsed with homogenizing medium. The medium contains 0.0375 M Tris-maleate pH 6.5, 0.5 M sucrose, 1% dextran, and 0.005 M $MgCl_2$. They are then homogenized for 30-60 sec in a Polytron ST20 operated at the slowest speed. About 20 ml of medium is used for each 10 g fresh weight of liver. The homogenate is squeezed through a single layer of Miracloth and centrifuged at 2000 x *g* for 15-30 min in a swinging-bucket rotor. A biphasic pellet forms, consisting of a pink bottom layer and a tan top. All of the top layer is removed and is gently resuspended in about 2 ml of the medium described above for each liver homogenized. This suspension is then layered over 3 vol of 1.25 M sucrose containing buffer, dextran, and $MgCl_2$ in the concentrations given above, and centrifuged at 100,000 x *g* for 20 min in a swinging bucket rotor.

The Golgi apparatus collect in a white band at the 0.5/1.25 M sucrose interface (density about 1.19), and contaminating particulate material sediments to the bottom of the tube. The density of these organelles has been reported to be 1.13 [21]. The white lipid layer at the top of the tube is drawn off by a touch with a piece of absorbent paper. The Golgi apparatus layer is removed by use of a pipette, along with as little of the sucrose solution as possible, and is mixed with an equal amount of cold distilled water. It is then sedimented at 10,000 x *g* for 10 min. The purity may be improved further by 3 or 4 resuspensions (in homogenizing medium) and recentrifugation. The yield from this technique is 5-10 mg of Golgi apparatus protein per 10 g of liver and the recovery is estimated to be as much as 70% of the Golgi apparatus in the cell. It has been reported [21] that this technique can be scaled up with the use of large rotors so that the livers of 10-12 rats can be processed in a single run of 3 hours or less to yield 50-100 mg Golgi apparatus protein. The process is shown in Fig. 2.

The original description of this technique called for resuspension of the Golgi apparatus in clarified rat-liver supernatant (homogenate centrifuged at 100,000 x *g* for 30-120 min) but in subsequent experiments in my hands this procedure has not resulted in significant improvement in the preservation of Golgi apparatus, and it does significantly increase the possibility of introducing extraneous material. It also makes protein recovery calculation difficult.

A similar procedure to the one described above has been employed by Schachter et al. [7] and by Wagner and Cytkin [23], who homogenized rat livers in a Potter-Elvjhem-type, Teflon-glass homogenizer (A. Thomas Co., size C). In the former case, the discontinuous sucrose gradient contained layers of 0.7 M, 1.3 M, and 1.7 M. Golgi apparatus membranes were reported to collect at the 0.7/1.3 M interface.

A novel technique has been developed by Ehrenreich et al. [24] for the isolation of Golgi membranes from ethanol-treated rats. This technique takes advantage of the change in density of the membranes caused by the accumulation of very-low-density lipoprotein particles in the Golgi apparatus. Fragmented Golgi cisternae accumulate at the 0.25/0.6 M interface in a discontinuous sucrose gradient. The purification and specific activity of the galactosyl transferase activity by this procedure is the highest yet reported (see Table 1). The cisternae are unstacked, however, and important structures or enzyme activities may be lost from such preparations.

A procedure similar to that described by Morre et al. [22] for rat liver has also been used to isolate Golgi apparatus from the mucopolysaccharide-screting gland of the snail (<u>Helix pomatia</u>) with excellent structural preservation by Ovtracht et al. [25]. These large Golgi complexes would appear to be a very interesting subject for structural and enzymatic studies.

TABLE 1

Comparison of Golgi Apparatus Isolation Methods

Tissue	Reference	Morphology preservation	Gal transferase specific activity mum/hr/mg	Enrichment over homog.	Yield
Rat liver	Morre et al.[6]	Fair	228 ± 11	95X	5-10 mg/10 g (1 liver)
Bovine liver	Fleischer et al.[5]	Poor	80	40X	18 mg/330 g
Snail	Ovtracht et al.[25]	Good	15 ± 5	6-25X	3.5 mg/4 g (25 snails)
Testis	Cunningham et al.[27]	Good	167 ± 12	12X	3-6 mg/9 g (6 testis)
Rat liver	Ehrenreich et al.[24]	Poor	200-250	55-110X	--

D. Bovine Liver

A technique for the isolation of Golgi apparatus from bovine liver has been reported by Fleischer et al. [5]. Liver tissue is trimmed of connective tissue and ground in a meat grinder. The mince is suspended in a 0.5 M sucrose solution containing 0.1 M sodium phosphate buffer, pH 7.2, and 1% dextran 500. It is then homogenized briefly in a Potter-Elvjhem homogenizer with a clearance of 0.025 in, driven by a motor at 1000 rpm. The pellet sedimenting between 8700 x _g_ for 30 min and 34,800 x _g_ for 30 min is resuspended in 42.7% sucrose and is loaded into a Spinco B 14 zonal rotor containing a step gradient of 16%, 23.9%, 34.5%, and 37.3% sucrose. After centrifugation for 45 min at 35,000 rpm, the rotor is unloaded by the pumping of heavy sucrose into the center and collection of 20-ml aliquots.

The yield of Golgi membranes from this technique is larger than from tissues of smaller animals (see Table 1) but is not commensurate with the volume of tissue and medium used. The initial report [5] of this technique indicates a lower specific activity and purification for galactosyl transferase than that reported for rat liver [6]. A more recent report [26] indicates a higher activity for the transferase but gives no data on purification or yield of this enzyme.

The morphological preservation obtained in zonal rotor centrifugation does not appear to be good, probably because of the long time required to load and unload the rotor under normal conditions, which results in deterioration of the organelles.

E. Rat Testis

A technique was recently developed by Cunningham et al. [27] for the isolation of Golgi apparatus from rat testis. This technique is a modification of the method of Morre et al. [22]. It yields Golgi apparatus with very good structural preservation by a fairly simple procedure. The specific activity of the galactosyl transferase compares favorably (see Table 1) with the enzymatic activity of rat liver Golgi apparatus, but the purification of this enzyme is much lower and contamination of the fraction is obvious (about 30 to 50%) in the electron microscope. This fraction is of interest, nevertheless, because of the complex organization of organelles and because they may be involved in the secretion of both the lysosomelike acrosome, which is stored inside the cell, and spiny coated vesicles (see Fig. 3), which are probably secreted from the cell.

Procedure: Adult male rats (200-300 g) are killed by cervical dislocation, the testes are excised, the tunica albugenia and testicular artery are removed, and the soft mass of seminiferous tubules are dropped into ice-cold medium containing 0.25 M sucrose, 0.05 M Tris-maleate buffer, pH 6.5, 1% dextran (Sigma 200C), and 5 mM $MgCl_2$. Two testes (weighing about 3-4 g) are homogenized in 10 ml of medium in a 50-ml centrifuge tube with a Polytran 10ST at the lowest possible speed (about 1500 rpm) for 20 sec. An equal volume of homogenizing solution is mixed in and the suspension is centrifuged at 2000 x *g* for 10 min in a swinging-bucket rotor (e.g., Sorvall HB4). A biphasic pellet forms which has a red-pink bottom layer and a soft, tan top. The cloudy supernatant is discarded and the entire tan layer is resuspended in about 30 ml of fresh medium containing 1.1 M sucrose, 0.05 M Tris-maleate buffer, pH 6.5, and 5 mM $MgCl_2$ and 1% dextran. This suspension is transferred to a clean centrifuge tube and 0.8 M sucrose is layered on top. The suspension is centrifuged at about 100,000 x *g* for 1 hr in a swinging bucket rotor (e.g., Spinco SW25.1 at 25,000 rpm).

The Golgi apparatus are buoyant at a density of approximately 1.13 and will form a fluffy or granular white layer at the 1.1/0.8 M interface. As much as possible of the 0.8 M sucrose layer is removed with a pipette and discarded. The Golgi apparatus layer is removed along with as little of the 1.1-M sucrose as possible, and diluted with an equal volume of ice-cold distilled water. The Golgi complexes are then sedimented at 10,000 x *g* for 10 min. The resulting pellet, which should contain about 1-2 mg protein for each rat used, should be smooth, firm, even, and white. The Golgi apparatus should be used immediately because deterioration is rapid. Within a few hours, even at 0°C, all structure is destroyed. The best morphological preservation occurs in preparations in which the Golgi apparatus layer in the density gradient has a granular appearance. In such a fraction, large areas occur which seem to be very pure Golgi apparatus, interspersed with areas of greater contamination. Sampling for morphological estimates of purity must therefore include a large number of areas. For examples of such a preparation see Figs. 1, 4, 5, and 8.

Discussion

1. Contamination

The most conspicuous contaminant of this preparation is variable-sized droplets of cytoplasm surrounded by plasma membrane and containing mitochondria, star-shaped aggregates of ER, senescent Golgi complexes, and ribosomes. The amount of this contaminant varies from a majority to perhaps 5% of the fraction (based on morphological observations). These droplets are probably derived from cytoplasm cast off in spermatogenesis. The reason for their variable presence is not clear but it appears to be some subtle variation in the homogenization. A modification of this procedure has been developed by Mollenhauer et al. [28] in which the homogenization is accomplished by passage of the seminiferous tubules through a Yeda press at 350-400 psi. This procedure is reported to give an equal structural preservation of the Golgi apparatus under more reproducible conditions and to eliminate cytoplasmic-droplet contamination.

Attempts to homogenize seminiferous tubules with a French press result in no recoverable Golgi apparatus, but passage through a 25-gauge syring needle yields tubular and vesicular membranes (see Fig. 6) that sediment like Golgi apparatus and that are entirely free of lysosomes, mitochondria, and cytoplasmic droplets.

In addition to the cytoplasmic droplets, the Golgi apparatus fraction isolated by this technique contains a considerable amount of lysosomes, lipid bodies, mitochondria, rough ER, and unidentifiable smooth-membrane vesicles (see Fig. 1). Some of the latter may be Golgi-apparatus derived. It would seem that the difference in size and density between the Golgi apparatus and these contaminating particles would be so great that further purification should be easy. This has not proven to be the case, however. Differential centrifugation or further density gradient centrifugation (either linear or discontinuous gradients) simply produce less of the fraction with an approximately constant level of contamination. This observation coupled with the granular nature of the Golgi apparatus layer in step or linear gradients may explain something of the effectiveness of the Polytron in releasing Golgi complexes.

When the Polytron is run at very low speed (or the tissue is chopped with razor blades), the cells are broken with a minimum of shear, and large areas of cytoplasm containing Golgi apparatus and neighboring organelles are released. These areas appear to aggregate, for some reason, and their sedimentation is established by the large Golgi apparatus. The morphology of the Golgi apparatus seems to be preserved as long as the cellular milieu is intact, but attempts to release contaminating particles by further homogenization or washing simply result in destruction of the Golgi apparatus.

Rodent testis is uniquely suited for Golgi-apparatus isolation. The seminiferous tubules contain very little connective tissue and are soft and easily broken. This is not true of the testes of larger animals such as bulls, resulting in an impediment to scaling up the process with larger testes. The spermatocyte and spermatid cells, which seem to break preferentially, contain very little ER. The density of the plasma membrane is increased because mitochondria become attached to it during spermiogenesis and this aggregate separates readily from the Golgi apparatus during centrifugation.

2. Homogenization Medium

a. Sucrose Concentration. Golgi apparatus may be isolated from rat testis in sucrose concentrations ranging from 0.25 to 1.0 M. Preservation is slightly better at higher sucrose concentrations, and if the molarity is high enough (1.0 M), the Golgi apparatus can be collected at the top of the tube after a single centrifugation, eliminating the initial sedimentation and resuspension in dense sucrose. The preservation of the cytoplasmic droplets is also markedly enhanced, however, by high sucrose concentrations. Homogenizing in 0.25 M sucrose represents a good compromise between Golgi apparatus preservation and purity.

b. Buffer. There seems to be some beneficial effect of Tris-maleate buffer on Golgi-apparatus isolation. The pH can be varied between pH 6.5 and 8.0 without much visible effect on the isolated organelles, but Tris-maleate seems to work better than Tris-HCl or phosphate.

c. Dextran. The compound dextran also has an unexplained beneficial effect on Golgi-apparatus isolation. High molecular weights (over 100,000) seem best and concentrations from 1 to 5% are effective.

d. Magnesium. Divalent cations definitely produce better preservation and purification of Golgi apparatus. Again, it is difficult to explain the basis of this empirical fact, but Mg^{++} is more effective than Ca^{++}.

e. Albumin. Addition of 0.1 to 1.0% albumin (e.g., bovine serum albumin) sometimes seems to improve preservation of the isolated Golgi apparatus, but the effects are subtle and its use does not usually seem justified.

f. Sulfhydryl Agents. Mercaptoethanol increases the activity of the galactosyl transferase but has no apparent effect on structural preservation.

3. Homogenization Time and Speed

Increasing the length of homogenization time from 20 to 60 sec, or the speed of the Polytron from 1000 to 5000 rpm, increases the number of cells broken but does not appreciably affect the yield of Golgi apparatus. This implies that an equilibrium in the number of free organelles is quickly reached and that there is destruction of free Golgi apparatus concomitant with the release of additional ones. It is much better to do several short homogenizations, collecting the free Golgi apparatus by centrifugation each time. Very high Polytron speeds (10,000-20,000 rpm) destroy the Golgi apparatus completely; at the low speeds used in these procedures, however, sonic energy and high shear (for which the Polytron was designed) are not produced.

The concentration of tissue in the homogenizing fluid does seem as important as does the geometry of the homogenizing vessel, perhaps because the geometry controls the number of times an individual Golgi apparatus passes through the blades. The best results with testis have been obtained with a 30% homogenate (w/v) in a centrifuge tube only slightly larger than the Polytron shaft. The suspension should be diluted somewhat before centrifugation, however, because the concentration of the tissue in the homogenate alters the density of the solution and changes the sedimentation. A 15-20% suspension is appropriate for centrifugation.

4. Centrifugation Techniques

It would seem that the first low-speed centrifugation (2000 x g for 5 min) in this technique could be lengthened since the supernatant solution is very turbid. Very few Golgi apparatus are contained in the supernatant, however, and only contamination is added by longer or harder centrifugation.

Discontinuous sucrose gradients are preferable in this procedure because they are much easier to make than are linear gradients and because they can handle much more material. If the Golgi-apparatus suspension is placed on the top of either a step or linear gradient, care must be taken that the concentration of Golgi apparatus is not too great or the gradient will overload and contamination will occur. By resuspending the Golgi apparatus in a sucrose solution of high density and sedimenting them toward the center of rotation, more material can be used, and organelles pass each other in solution rather than filter through a thick layer at an interface.

5. Method of Sacrificing Animal

The testis appears to be much less sensitive to the condition of the animal than does the liver. It does not seem to affect the Golgi-apparatus preparation whether the animal is fed or starved before death. It seems desirable, however, to avoid the use of anesthetics or organic solvents such as chloroform or ether which very likely change membrane structures. Suffocation of the animal in CO_2 atmosphere (a container with dry ice and water underneath a screen on which the animal is placed) seems a quick and relatively painless death and does not affect Golgi-apparatus isolation.

III. CRITERIA FOR ASSESSING ISOLATION METHODS

A. Morphology

The Golgi apparatus was originally recognized as a morphological feature of the cell, and the first attempts to isolate this structure depended necessarily on morphological observations for the assessment of purification and preservation of the organelles. More data are available now on the enzymatic and chemical characteristics of this complex, but structural studies, especially at the electron microscope level, are still indispensable. The most reliable technique for observing Golgi-apparatus morphology is through fixation, embedding, and sectioning. The procedures are time consuming and tedious, and they preclude exploratory experiments in which one might assess the effectiveness of a particular treatment before continuing to subsequent steps. This technique does, however, reveal a more easily interpretable view of both the state of the Golgi apparatus and possible contaminating structures than does any other available method.

1. Procedure for Fixation and Sectioning

a. Collection of Particles. A suitable aliquot of the fraction to be examined is transferred to a flexible plastic centrifuge tube to make a pellet of appropriate thickness for fixation, and is diluted with sucrose, if necessary, to fill the tube. For example, a 0.5-ml aliquot of a suspension that contains about 1 mg protein/ml should make a pellet about 0.5 mm thick in a Beckman #305528 (3/16 x 1 5/8 in.) tube. These small tubes are convenient because they require little sample and can be used, with appropriate adapters, in either a Spinco (adapter #305527 for SW50.1 head) or Sorvall (adapter #423 for either SS34 or HB4 head) centrifuge. The pellet must be thin enough for the fixative to penetrate adequately, and yet thick enough so that sections will contain a valid representation of the contents of the fraction. The tube should be spun at a speed sufficient to make a firm pellet which will withstand subsequent handling, but not so tightly packed as to make resolution of structures and interpretation difficult.

b. Fixation. The fixation must preserve all the structures present in the fraction. Care must also be taken to retain the orientation of the pellet since, in centrifugation, the organelles will layer to a surprising extent with the more dense and larger structures in definite strata at the bottom of the pellet and lighter structures at the top, regardless of their original positions in the tube. It is, therefore, wise to leave the pellet attached to the bottom of the centrifuge tube during the initial steps in preparation so that the pellet will not break up and so that the curvature of the bottom will aid in orientation during sectioning. The bottom of the flexible cellulose-nitrate tube is easily sliced off with a razor blade, and if the pellet is thicker than 0.5 mm it can be divided further into halves or quarters to aid penetration of the fixative. The small piece of plastic adhering to the bottom of the pellet makes it possible to decant fluids easily in the early stages of fixation and dehydration, and it dissolves in acetone or propylene oxide before plastic infiltration.

A good fixative for isolated fractions is 2.5% glutaraldehyde in 0.1 collidine or phosphate buffer, pH 7.0-7.4. The glutaraldehyde should be stabilized with Amberlite, stored in ampules, or purified with charcoal and/or barium carbonate before use [29]. Fixation time depends on the thickness of the pellet, but 2 hr at 0°-4°C is usually adequate, and up to 12 hr at 4°C does no apparent damage.

c. Staining. Contrast is introduced by staining of the pellets with osmium tetroxide (OsO_4) after thorough washing to remove the glutaraldehyde. The wash solution should be the same buffer used for fixation, with the addition of 0.2 M sucrose to approximate the osmolarity of the glutaraldehyde. A 1-2% OsO_4 solution is prepared in the same buffer (osmolarity does not seem to be as important in the staining solution), and the pellets are stained for 1-4 hr at 4°C. The pellets should be blackened uniformly.

d. Dehydration and Embedding. These steps are the same as those in the preparative technique of any electron microscope [30]. Dehydration may be carried out in a graded acetone series (25-50-75-100%), and infiltration of any of the epoxy-resin mixtures is usually accomplished fairly quickly because of the lack of cell walls, etc. Thin sections are cut and poststained with uranyl acetate and lead citrate [30].

Discussion: Figure 1 shows a representative section through a Golgi-apparatus preparation from rat testis. The organelles round up after extraction from the cell, but the organization of the cisternal stacks are seen to be well preserved. An anastomosing system of tubules and vesicular profiles (mainly cross-sectioned tubules) can be seen around the periphery of the stacks. Note the spiny-coated vesicles [14] in their typical orientation on the maturing face of the apparatus. Mitochondria, lysosomes, lipid bodies, and rough endoplasmic reticulum (RER) can also be identified. Many smooth-membrane vesicles can also be seen but their origin cannot be ascertained by this method.

Golgi-apparatus membranes accumulate osmium deposits under long staining times at high temperatures [31], but this reaction is not specific enough to identify Golgi-apparatus membranes in heterogeneous fractions, since lysosomes, ER, and even mitochondria also stain to some extent under these conditions.

2. Negative-Contrast Staining

The negative-contrast staining process is extremely useful as a rapid, simple technique for assaying morphological preservation and purity [32,33]. It can be used while the isolation procedure is in progress, to indicate the efficiency of previous steps and the necessity for further manipulation. The images obtained by this procedure must be interpreted with caution, however, since large amounts of protein are extracted by negative stains [34], and morphological alterations are not unlikely. Furthermore, high concentrations of sucrose, buffer, and Dextran, which are necessary in the isolation medium to preserve structure, will obscure the negative-contrast image. It is necessary to wash out these compounds or to dilute them to a low level, but this hypotonic treatment can be shown to cause extensive vesiculation or tubulation of the Golgi apparatus [35]. Certain contaminants, furthermore, are not visualized at all by this technique. For instance, the cytoplasmic droplets that are a major contaminant in Golgi-apparatus fraction from the testis (described below) are not detected in negative staining and must be assayed by fixation and sectioning.

Procedure: A 2% solution of uranyl acetate or ammonium molybdate in H_2O, or potassium phosphotungstate (PTA) neutralized to pH 6.8-7.0 with KOH, will work as a negative stain for visualizing isolated organelles. The PTA gives the sharpest resolution and most pleasing images because of its smaller grain size and better spreading properties. The latter probably results from the extraction of protein from the membranes by PTA at neutral pH.

The particles are collected by sedimentation and diluted with water to a protein concentration of about 1 mg/ml. This concentration will give a barely visible turbidity in the tip of a disposable Pasteur pipette. A drop is placed on a carbon-coated, collodian, or Formvar-coated grid, and all but a thin film is drawn off with filter paper. Before the thin film of liquid dries, a drop of negative-staining solution is added and the excess is drawn off again with filter paper. The grid is air dried and may be observed immediately or stored for several weeks. For a further discussion of negative staining, see Ref. [32].

Resuspension in water will usually dilute any sucrose or buffer present in the fraction so that it will not obscure the image. Sometimes a suspension that contains too much sucrose or buffer can be washed on the grid by the addition of several successive drops of stain or distilled water and a blotting between drops. If the surface tension is so high that the suspension will not wet the grid and form a thin film, a small amount of bovine-serum albumin is added (by trial and error) to the suspension. Alternatively, the surface of the carbon coat can be activated by exposure to slow-discharge in a vacuum evaporator for a few seconds.

A typical rat spermatocyte Golgi apparatus is shown in negative contrast in Fig. 4. The characteristic anastomosing network of tubules attached to fenestrated plates is a distinctive feature of all Golgi complexes observed so far. The tubules have an irregular width ranging from 300 to 500 Å, and are characteristically crooked and branched. They also absorb a slight amount of stain so that the resulting contrast is a combination of positive and negative stains. Usually the fenestrate and solid cisternae in the center of the complex are obscured by the tubules around the periphery of the cisternae.

Discussion: The presence of tubules per se is not a sufficient indication of Golgi-apparatus membranes. It is very likely that smooth endoplasmic reticulum and plasma membrane also break up into tubules after extraction. Figure 7 shows a cluster of tubule membranes not derived from a Golgi apparatus. These tubules have different dimensions than those of the Golgi apparatus, do not anastomose, and are much more stain repellant. For a further discussion of membrane identification in negative staining, see Cunningham and Crane [33].

When the cisternal stack slides apart, the fenestrated plates in the center of the apparatus become more apparent.

Figure 5 shows an example of such cisternae. This is probably an indication that the preparation is beinning to deteriorate.

The fidelity of sampling of particulate fractions with this technique may be a problem. Most large membranous organelles will be preserved, but there is a high probability that small proteinaceous particles such as ribosomes, microtubules, collagen fibers, etc., will be washed off the grid or dissolved in the stain. The only solution to this problem is to carefully compare the results obtained by several techniques and to be aware of possible artifacts when making interpretations.

3. Freeze Fracturing

This technique has the advantages of being more rapid than fixation and sectioning, avoiding the osmotic damage and extraction that occurs in negative staining, and revealing membrane surfaces and the internal structure of membranes.

Procedure: The organelles must be protected against freezing damage; therefore a suspension (in sucrose) is briefly fixed by the addition of glutaraldehyde to make a 1% solution (final). After approximately 15-30 min at 0°C the glutaraldehyde is washed out by centrifugation and glycerol is added slowly with stirring (over a 15-min period). The organelles are then sedimented again at 10,000 x *g* for 10 min. Small chunks of the gelatinous pellet are scooped up, placed on copper or gold disks, and quickly frozen in Freon cooled to near its freezing point with liquid nitrogen. The sample is then transferred to a freeze-etch machine, fractured, shadowed with carbon and platinum, and then replicated with a thick carbon layer under vacuum [36,37]. The organic material is cleaned with sodium hypochlorite (bleach) and chromic acid. The replica is washed thoroughly with water, picked up on a grid, and observed in the electron microscope.

Discussion: Figure 8 shows a fractured surface of a Golgi-apparatus preparation in which five cross-fractured Golgi complexes can be seen. When the fraction is glutaraldehyde fixed and glycerol infiltrated before freezing, the stacking of the cisternae is well preserved and the general morphology of the complex is much the same as in plastic sections (see Fig. 1). Freeze fracturing adds additional information, not available by other techniques, as to the condition of the membranes. The plane of fracture in the frozen sample follows areas rich in hydrophobic bonds (such as the interior of membranes), which are neutralized by the binding of water in ice crystals. Thus the interior structure of membranes is revealed by this technique and many subtle structural features may be distinguished.

In intact tissues [38] the fracture surface of Golgi-apparatus membranes is thickly studded with small (50 Å), evenly spaced granules (presumably protein molecules). In isolated fractions these granules appear to be sensitive to the state of preservation of the membranes. In preparations in which the membranes have begun to deteriorate [35], the granules aggregate into patches or into a reticulate pattern on the fractured face. In many areas (such as the membranes shown in Fig. 8), the fracture face may be completely devoid of granules, although the gross morphology of the Golgi apparatus looks normal. We interpret this as alteration in the composition or organization of membrane proteins (probably by lytic enzyme activity) that is not revealed by other techniques but that is probably very important in interpretations of Golgi-apparatus morphology and function.

B. Enzymatic Assays

The first successful isolation of the Golgi apparatus [17] showed a concentration of acid phosphatase activity and lipid phosphorus in the Golgi-apparatus fraction. However, recent results with other Golgi fractions [39] suggest that these characteristics may not clearly distinguish Golgi-apparatus membranes from other microsomal fragments. Thiamine pyrophosphatase activity has been shown cytochemically [9] to be localized in the Golgi apparatus, but the enrichment of this enzyme in Golgi-apparatus fractions from rat liver is not encouraging for use of this enzyme as a marker for the Golgi apparatus. The best enzyme activity demonstrated to date as an indication of Golgi-apparatus purification is UDP-galactose:n-acetylglucosamine galactosyltransferase [6,7]. This enzyme is assayed by the transfer of radioactively labeled ^{14}C-galactose from USP to n-acetyl glucosamine and by the difference in binding of these compounds to Biogel (Dowex) columns.

Procedure (from Morre et al. [6] and Cunningham et al. [27]): Triton X-100 is added to the organelle suspension to make a final concentration of 1%, an amount that disperses the membranes and releases enzyme activity. Duplicate aliquots containing 0.01-0.2 mg protein are incubated for 10-30 min at 37°C with gentle shaking in a 0.3-ml reaction mixture containing 2 μM Tris-HCl (pH 7.5), 1 μM $MgCl_2$, 0.5 μM mercaptoethanol, and 61 nM uridinediphosphate galactose (UDP-Gal, containing 9 pCi UDP-Gal-^{14}C from New England Nuclear, specific activity 252 Ci/mole). One reaction of each duplicate set contains 90 nM of n-acetylglucosamine (NAG) to measure transfer activity, while the other sample has no NAG to control for nonspecific hydrolysis. The UDP-Gal solutions should be freshly made and should be checked for the presence of free galactose before use. The reaction is stopped by placement of the mixture on an anion exchange column consisting of about 1 ml of AG_1-X_2 Biogel resin (Cl^- form, 200-400 mesh, Bio-Rad Laboratories, Richmond, Cal.) held in a disposable Pasteur pipette with a glass wool plug. The reaction products (n-acetylgalactosamine and free galactose) are eluted from the column with 3 x 0.4 ml washes into a scintillation vial, while the non-reacted UDP-Gal remains bound to the resin. A water-miscable fluor (e.g., Aquafluor, New England Nuclear) is added to the vials, and radioactivity is counted in a liquid scintillation counter. Protein is determined by the Lowry method [40] and activity is usually expressed as mumoles (or nM) Gal transferred per hr per mg protein. Other methods for determining enzyme activities useful for the determination of contamination in the Golgi apparatus are given in Cheetham [39] and Morre [21].

Discussion: The galactosyl transferase is probably not the only complex carbohydrate synthetic activity localized in the Golgi apparatus. Schachter et al. [7] have demonstrated sialyl-transferase and n-acetylglucosaminyltransferase activity in a Golgi-rich fraction from a rat liver. They also showed that activities for galactosyl transfer are higher when a glycoprotein acceptor, rather than NAG, is used. This probably indicates that a growing glycopeptide chain is the natural acceptor in the Golgi apparatus. Their assays included 2(n-morpholino)ethene sulfonate buffer and dithiothreitol rather than Tris and mercaptoethanol; since they did not present specific activities, however, it is difficult to determine whether these reagents are better than the ones described here.

Table 1 shows some previously reported enzyme activities for Golgi-apparatus fractions isolated from a variety of cell types.

IV. CONCLUSIONS

In designing an isolation procedure for the extraction of Golgi apparatus, it appears that the following points should be considered:

1. The Golgi apparatus are highly sensitive to lytic enzymes in the homogenate (and perhaps in the cisternae or vesicles themselves), and to minimize damage the procedure should be carried out as quickly as possible with all solutions kept between 0° and 4°C.

2. Golgi apparatus are sensitive to osmotic disruption and excessive homogenization. The homogenizer should be as gentle as possible (razor blade chopping, Polytron, or very loose Potter-Elvejhem) and the sucrose concentration should be at least 0.25-0.5 M.

3. Magnesium, Tris-maleate buffer, high-molecular-weight Dextran, and albumin appear to improve preservation and/or purification of Golgi apparatus. Sulfhydryl agents improve enzyme activities.

4. Intact Golgi apparatus are likely to sediment at low speeds (1000 to 2000 x *g* for 10 min) in 0.25-0.5 M sucrose, and most Golgi apparatus isolated so far appear to have a density around 1.12-1.14 unless loaded with low density lipid, so that density gradient techniques can be employed to separate them from other cellular organelles.

5. Morphological observation (especially fixation and sectioning) is the most reliable method for determining preservation and purity of isolated Golgi apparatus, but galactosyl transferase may be a universal and much more easily quantifiable marker for Golgi apparatus.

ACKNOWLEDGMENT

I would like to express my appreciation to Drs. L. A. Staehelin and Mary Bonneville for reading the manuscript of this article and making helpful suggestions. Some of the original observations presented here were made in collaboration with Dr. R. Rubin, Mr. R. Wilkins, and Mrs. V. Fonte. I would also like to thank Dr. Keith R. Porter and the Department of Molecular, Cellular and Developmental Biology, University of Colorado, for making facilities available for writing this article.

REFERENCES

1. W. H. Beams and R. G. Kessel, Intern. Rev. Cytol., 23, 209 (1968).

2. P. Favard, in Handbook of Molecular Cytology (A. Lima de Faria, ed.), North Holland, Amsterdam, 1970, Chap. 37.

3. H. H. Mollenhauer and D. J. Morre, Ann. Rev. Plant Physiol., 17, 27 (1966).

4. D. H. Northcote, Endeavor, 30, 26 (1971).

5. B. Fleischer, S. Fleischer, and H. Ozawa, J. Cell. Biol., 43, 59 (1969).

6. D. J. Morre, L. M. Merlin, and T. W. Keenan, Biophys. Biochem. Acta, 37, 813 (1969).

7. H. Schachter, I. Jabbal, R. L. Hudgin, L. Pinteric, E. J. McGuire, and S. Roseman, J. Biol. Chem., 245, 1090 (1970).

8. M. Neutra and C. P. LeBlond, J. Cell. Biol., 30, 137 (1966).

9. P. M. Novikoff, A. B. Novikoff, N. Quintana, and J. J. Hauw, J. Cell. Biol., 50, 859 (1971).

10. D. Zagury, J. W. Uhr, J. D. Jamieson, and G. E. Palade, J. Cell. Biol., 46, 52 (1970).

11. D. J. Morre, H. H. Mollenhauer, and C. E. Bracker, in Results and Problems in Cell Differentiation (T. Reinert and H. Ursprung, eds.), Vol. III, Springer-Verlag, Berlin, 1972.

12. A. Claude, J. Cell. Biol., 47, 745 (1970).

13. W. P. Cunningham, D. J. Morre, and H. H. Mollenhauer, J. Cell. Biol., 28, 169 (1966).

14. F. R. Susi, C. P. LeBlond, and Y. Clermont, Am. J. Anat., 130, 251 (1971).

15. E. Wisse, J. Ultrastruct. Res., 38, 528 (1972).

16. C. de Duve, J. Cell. Biol., 50, 20D (1971).

17. W. C. Schneider, A. J. Dalton, E. L. Kuff, and M. D. Felix, Nature, 172, 161 (1953).

18. E. L. Kuff and A. J. Dalton, in Subcellular Particles (T. Hayashi, ed.), Academic Press, New York, 1959, p. 114.

19. D. J. Morre, H. H. Mollenhauer, and J. E. Chambers, Exptl. Cell Res., 38, 672 (1965).

20. P. M. Ray, T. L. Shininger, and M. M. Ray, Proc. Natl. Acad. Sci., 64, 605 (1969).

21. D. J. Morre, in Methods in Enzymology (W. B. Jacoby, ed.), Vol. XXII, Academic Press, New York, 1971, p. 130.

22. D. J. Morre, R. L. Hamilton, H. H. Mollenhauer, R. W. Mahley, W. P. Cunningham, R. D. Cheetam, and L. S. LeQuire, J. Cell Biol., 44, 484 (1970).

23. R. R. Wagner and M. A. Cynkin, Biophys. Biochem. Res. Commun., 35, 139 (1969).

24. J. H. Ehrenreich, J. J. Bergeron, and G. E. Palade, J. Cell. Biol., 47, 55a (1970).

25. L. Ovtracht, D. J. Morre, and M. Merlin, J. Microscopie, 8, 989 (1969).

26. L. M. G. Van Golde, B. Fleischer, and S. Fleischer, Biophys. Biochem. Acta, 249, 318 (1971).

27. W. P. Cunningham, H. H. Mollenhauer, and S. E. Nyquist, J. Cell. Biol., 51, 273 (1971).

28. H. H. Mollenhauer, S. E. Nyquist, and K. Acuff, Biochem. Prep., in press (1972).

29. E. g., Glutaraldehyde stabilized with amberlyst, Serva Feinbiochemica, Heidleburg. See also, D. D. Sabatini, K. Bensch, and R. J. Barnett, J. Cell. Biol., 17, 19 (1963).

30. A. M. Glauert, in Techniques for Electron Microscopy (D. H. Kay, ed.), 2nd ed., F. A. Davis, Philadelphia, 1965, p. 166.

31. D. S. Friend and M. J. Murray, Am. J. Anat., 117, 135 (1965).

32. R. W. Horne, in Techniques for Electron Microscopy (D. H. Kay, ed.), 2nd ed., F. A. Davis, Philadelphia, 1965, p. 328.

33. W. P. Cunningham, J. W. Stiles, and F. L. Crane, Exptl. Cell Res., 40, 171 (1965).

34. W. P. Cunningham and R. Rubin, J. Cell Biol., 57, 601 (1972).

35. W. P. Cunningham, L. A. Staehelin, R. Wilkins, and M. Bonneville, in preparation (1972).

36. S. Bullivant, in Some Biological Techniques in Electron Microscopy (D. F. Parsons, ed.), Academic Press, New York, 1970, p. 101.

37. H. Moor and K. Mullenthaler, J. Cell Biol., 17, 609 (1963).

38. L. A. Staehelin and O. Kiermayer, J. Cell Sci., 7, 787 (1970)

39. R. D. Cheetham, D. J. Moore, and W. N. Yunghans, J. Cell. Biol., 44, 492 (1970).

40. O. H. Lowry, N. J. Rosebrough, A. L. Farr, and P. Randal, J. Biol. Chem., 193, 265 (1951).

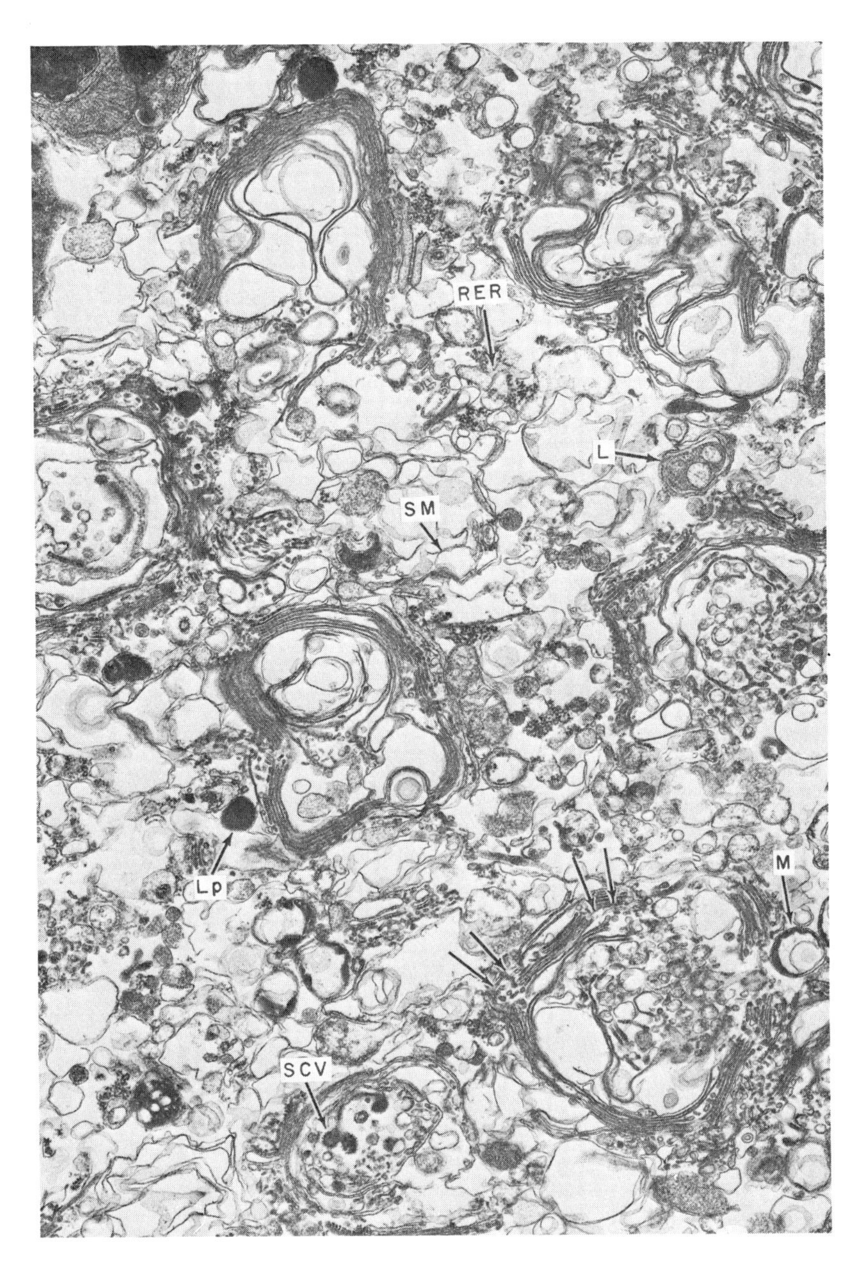

FIG. 1. Thin section through a pellet of Golgi apparatus isolated from rat testis. Note tubular interconnections between cisternal stacks (dictyosomes) in apparatus at lower right (double arrows). M = mitochondria, L = lysosome, RER = rough endoplasmic reticulum, Lp = lipid droplet, SM = smooth membrane (possibly of Golgi origin), SCV = spiny-coated vesicle. 11,000 x.

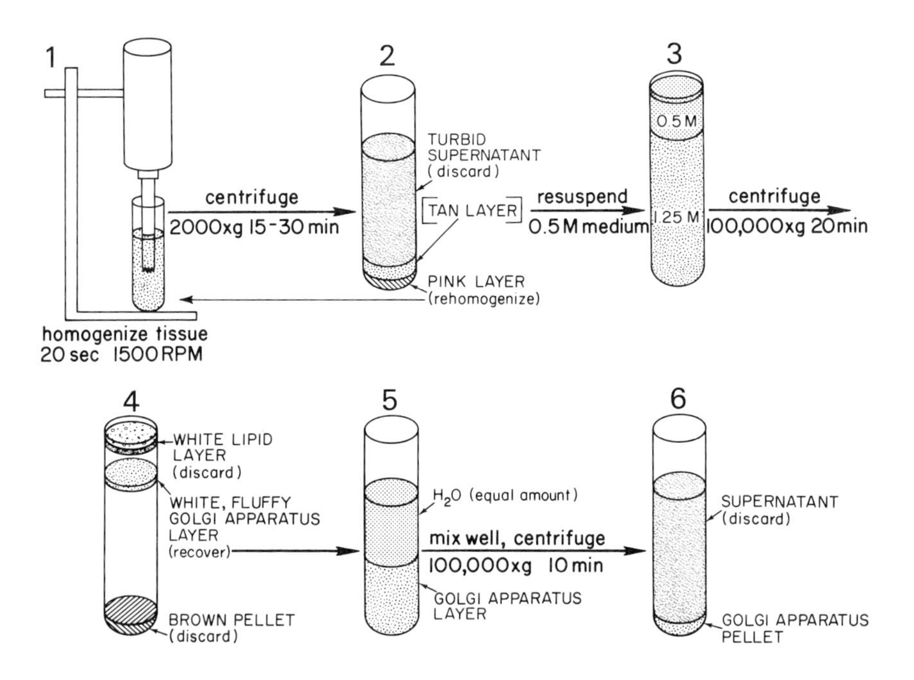

FIG. 2. Flow chart for rat liver Golgi apparatus isolation. Centrifugal forces given as average for tube.

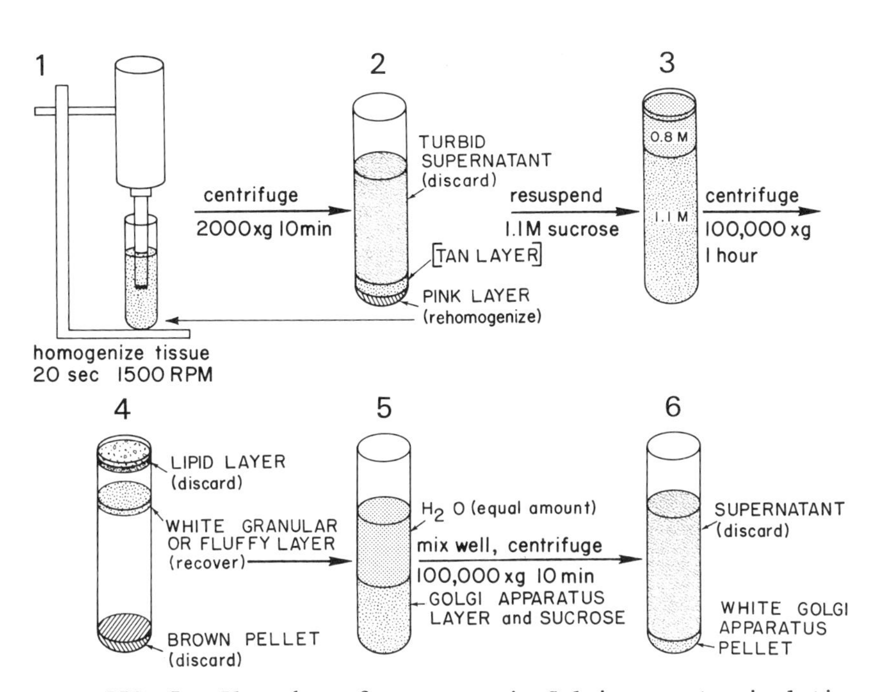

FIG. 3. Flow chart for rat testis Golgi apparatus isolation.

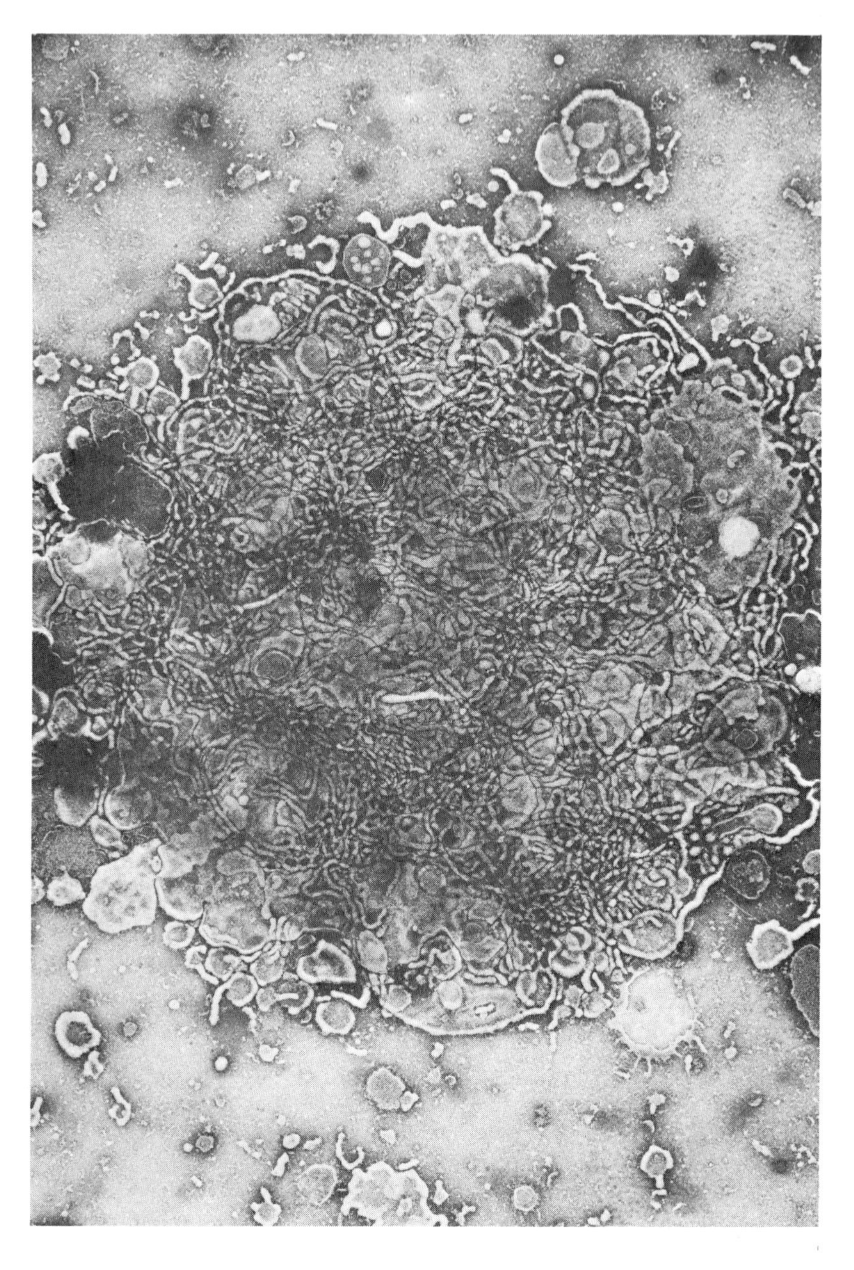

FIG. 4. Negative-contrast image of isolated rat testis Golgi apparatus, stained with neutral phosphotungstate. The peripheral tubules and the fenestrate upper cisternae predominate in this small, stacked Golgi apparatus. 40,000 x.

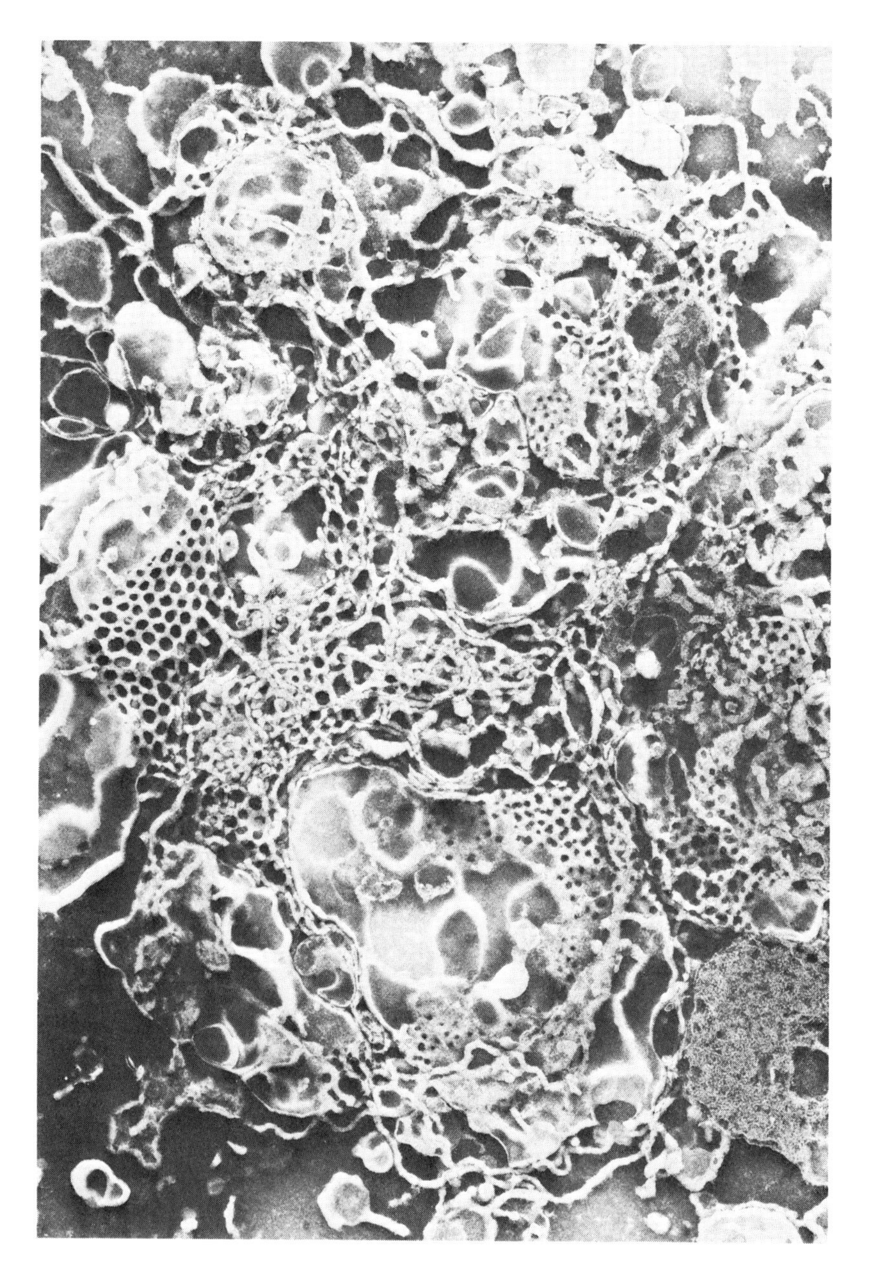

FIG. 5. Negative-contrast image of unstacked, isolated rat-testis Golgi apparatus. Note partially fenestrate cisternae. 47,000 x.

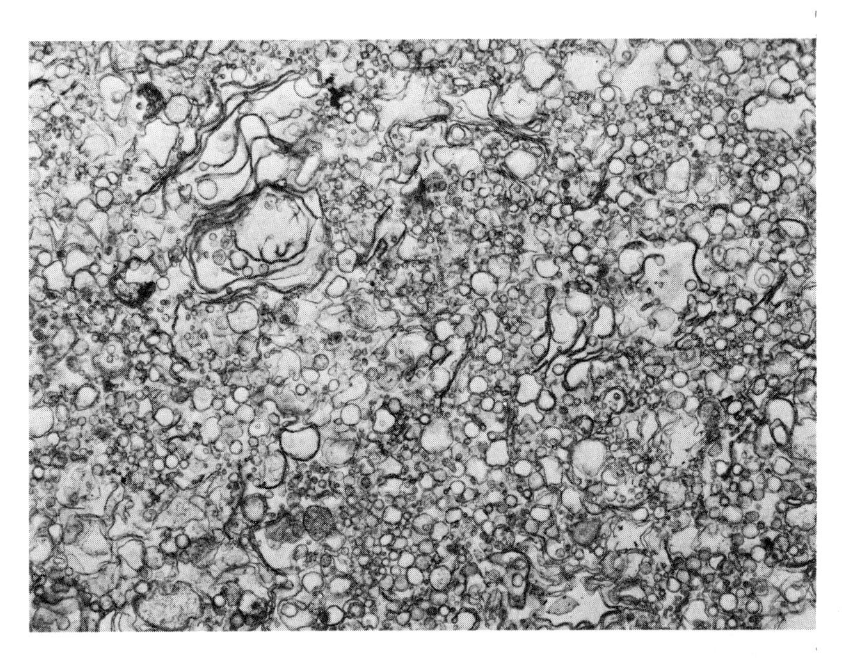

FIG. 6. Thin section through a pellet of smooth membranes isolated from rat testis by passage of seminiferous tubules through a syringe. Fraction buoyant at density of 1.13. Note absence of mitochondria, ribosomes, etc. 16,000 x.

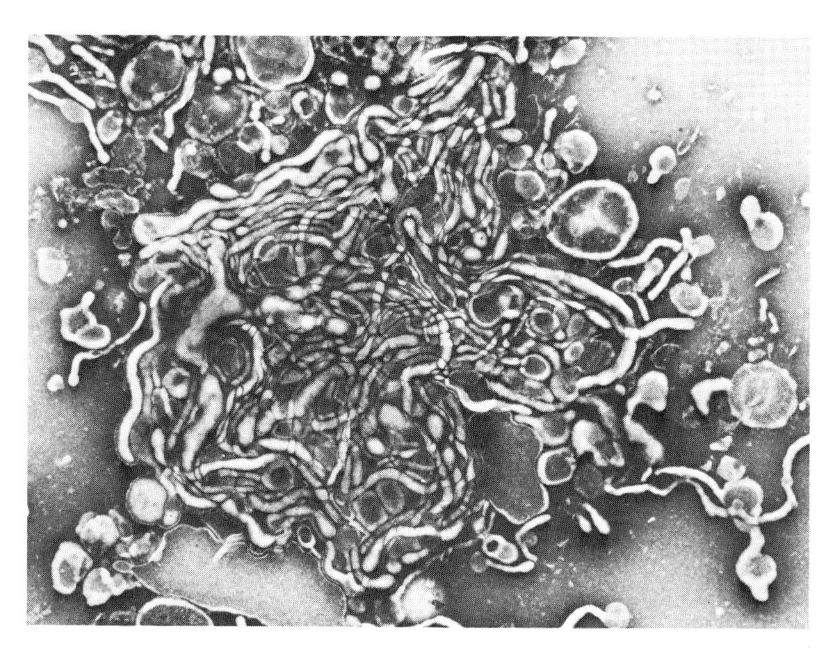

FIG. 7. Negative-contrast image of tubular membranes isolated from rat testis. Note differences between these membranes and those in Figs. 4 and 5. At high magnification these membranes can be seen to have a fringe of 50 Å particles along their edges (see Ref. [33]). 36,000 x.

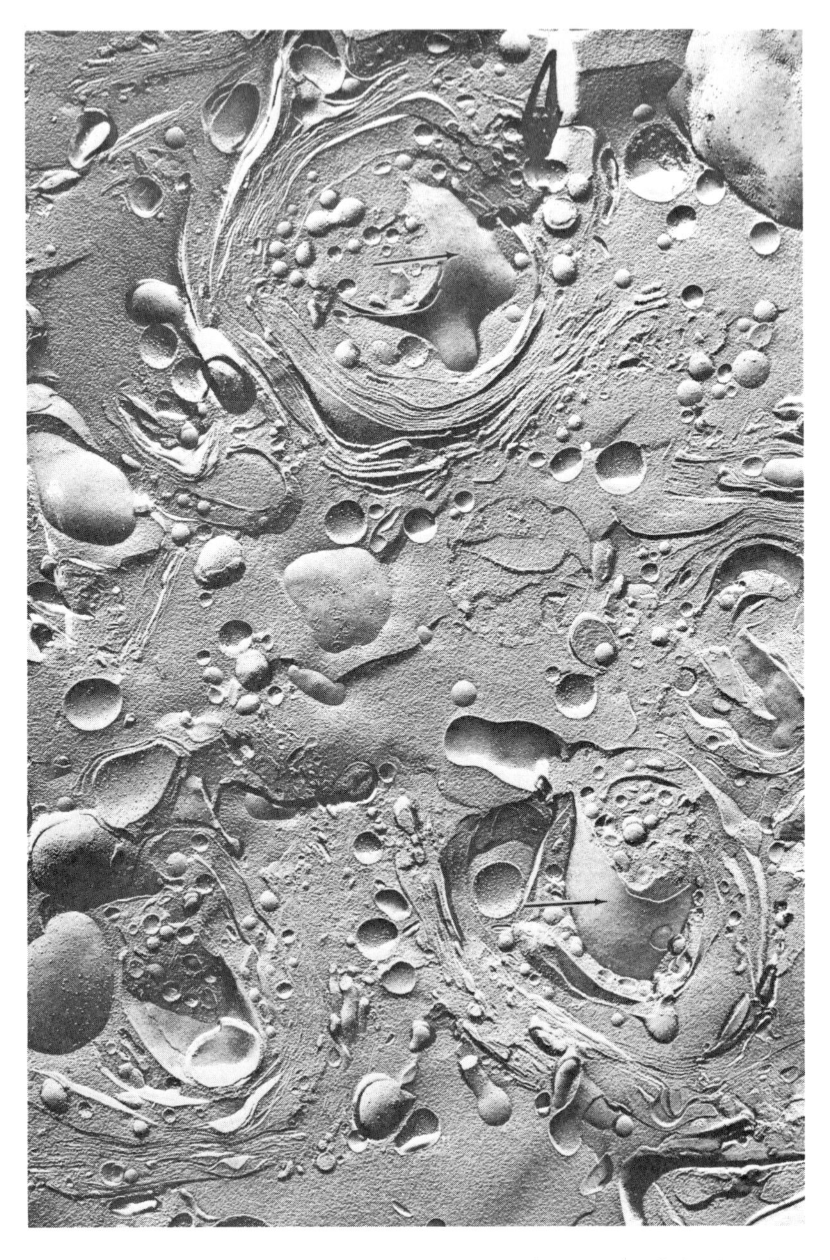

FIG. 8. Replica of frozen fractured pellet of isolated rat-testis Golgi apparatus. Note smooth-membrane fracture face (arrow). No etching. (Micrograph courtesy of Dr. L. A. Staehelin.) 16,000 x.

Chapter 6

ZONAL CENTRIFUGATION

C. A. Price

Department of Biochemistry and Microbiology
Rutgers University
New Brunswick, New Jersey

I. INTRODUCTION

Zonal rotors are hollow, closed cylinders into or out of which one can pump density gradients, sample suspensions, or separated particle fractions. We can understand what zonal rotors are by appreciating the reason for their invention: N. G. Anderson was at once fascinated by density-gradient centrifugation and dissatisfied with swinging-bucket rotors. He knew (see Ref. 1) that in principle density-gradient centrifugation was vastly superior to differential centrifugation for the separation of subcellular particles, but the principal vehicles for density-gradient centrifugation, swinging-bucket rotors, had very limited capacities and suffered from "wall effects," the tendency of particles to strike the walls of parallel-sided tubes, where their sedimentation would be greatly altered. Anderson [2] eliminated both problems with one stroke by extending the lateral walls of a tube through 360° (Fig. 1). A drawing of a typical zonal rotor is shown in Fig. 2. A comparison of the speeds and volumes of swinging-bucket and zonal rotors is given in Tables 1 and 2.

TABLE 1

Characteristics of Some Swinging-Bucket Rotors Commonly Used for Density-Gradient Centrifugation

Manufacturer and type	Place x volume	Maximum speed		g_{max}	r_{max} (cm)	$\Delta\ell$ (cm)
		rpm	ω^2			
Sorvall HL 8	8 x 100	3,700	1.50×10^5	3,720	24.5	10.0
	16 x 50	3,400	1.27×10^5	3,080	24.0	12.3
IEC 269	8 x 100	3,150	1.09×10^5	2,780	24.4	12.3
	8 x 50	4,100	1.84×10^5	3,720	19.8	9.3
IEC 253	12 x 50	3,000	1.27×10^5	2,400	22.7	9.3
	12 x 15	3,300	1.19×10^5	2,800	23.2	8.3
MSE 59591	3 x 65	23,500	6.06×10^6	100,000	16.2	9.8
Spinco SW 25.2	3 x 60	25,000	6.85×10^6	106,900	15.4	8.9
Sorvall HB-4	4 x 50	13,000	1.85×10^6	27,500	14.6	9.9
IEC SB-110	6 x 40	25,000	6.85×10^6	110,100	15.8	8.9
IEC 940	4 x 40	12,000	1.58×10^6	20,800	12.9	8.9
Spinco SW 27	6 x 38.5	27,000	8.00×10^6	131,000	16.2	8.9
Spinco SW 25.1	3 x 34	25,000	6.85×10^6	90,000	13.0	7.6

TABLE 1 cont'd.

Manufacturer and type	Place x volume	Maximum speed		g_{max}	r_{max} (cm)	$\Delta\ell$ (cm)
		rpm	ω^2			
MSE 59590	3 x 23	30,000	9.87×10^6	129,000	12.9	7.0
Spinco 27.1	6 x 17	27,000	8.00×10^6	135,000	16.7	10.2
IEC SB-283	6 x 12	41,000	1.84×10^7	283,100	15.1	9.6
IEC SB-206	6 x 12	35,000	1.34×10^7	206,000	15.1	9.6
Spinco SW 40 Ti	6 x 14	40,000	1.75×10^7	284,000	16.0	9.5
Spinco SW 36	4 x 13.5	36,000	1.42×10^7	193,000	13.4	7.6
Spinco SW 41 Ti	6 x 13	41,000	1.84×10^7	286,000	15.3	8.9
Spinco SW 50.1	6 x 5	50,000	2.74×10^7	300,000	10.8	5.1
Spinco SW 65L Ti	3 x 5	65,000	4.63×10^7	420,000	8.9	5.1
Spinco SW 50L	3 x 5	50,000	2.74×10^7	274,000	9.9	5.1
MSE 59587	3 x 5	65,000	4.63×10^7	420,000	8.9	5.1

MSE 59589	3 x 5	40,000	1.75×10^7	178,000	9.9	5.1
IEC SB-405	6 x 4.2	60,000	3.95×10^7	405,900	10.1	5.1
Spinco SW 56 Ti	6 x 4	56,000	3.44×10^7	408,000	11.7	6.0
MSE 59588	3 x 3	50,000	2.74×10^7	261,000	9.3	5.2

TABLE 2

Characteristics of Current, Batch-Type Zonal Rotors

Manufacturer designation	Nearest Oak Ridge designation	Maximum speed		Core configuration	Gradient volume in cc	Radius at edge in cm	g_{max}	Special characteristics
		rpm	ω^2					
IEC A-12	A-XII	4,600	2.32×10^5	Center loading	1300	18	4,206	Transparent end caps, largest radius
MSE A	A-XII	5,000	2.74×10^5	Center loading	1300	18	4,206	Transparent end caps, largest radius
IEC Z-8	--	8,000	7.02×10^5	Center loading,	1250	12	9,000	Transparent end caps
				edge loading	1115	11	7,800	Transparent end caps; MACS option
MSE HS	--	10,000	1.10×10^5	Center loading	695	10.3	11,400	Transparent end caps
Sorvall SZ-14	--	20,000	4.39×10^6	Reorienting gradient	1373	9.5	42,600	Sample normally loaded while spinning, but unloaded at rest; gradient reorients slowly during deceleration.

Spinco JCF-Z	--	20,000	4.39×10^6	Center loading	1900	8.9	40,000	
IEC B-29	B-XV	35,000	1.34×10^7	Center loading	1670	8.9	121,100	
	B-XXIX	35,000	1.34×10^7	Edge loading	1480	8.5	116,000	Sample may be loaded or unloaded at the center or the edge. MACS option.
MSE B-XV Ti	B-XV	35,000	1.34×10^7	Center loading	1670	8.9	122,000	
MSE B-XV Al	B-XV	25,000	6.85×10^6	Center loading	1670	8.9	62,000	
Spinco Ti 15	B-XV	32,000	1.12×10^7	Center loading	1675	8.9	101,300	
	B-XXIX	32,000	1.12×10^7	Edge loading	1350	8.4	95,600	Sample may be loaded or unloaded at the center or the edge.
Spinco Al 15	B-XV	22,000	5.31×10^6	Center loading	1675	8.9	48,100	
	B-XXIX	22,000	5.31×10^6	Edge loading	1350	8.4	45,200	Sample may be loaded or unloaded at the center or the edge.
Spinco B-4	B-IV	40,000	1.75×10^7	Center loading	1750	4.9	86,000	Requires special L-4 drive unit.

TABLE 2 cont'd.

Manufacturer designation	Nearest Oak Ridge designation	Maximum speed rpm	Maximum speed ω^2	Core configuration	Gradient volume in cc	Radius at edge in cm	g_{max}	Special characteristics
IEC B-30	B-XIV	50,000	2.74×10^7	Center loading	659	6.7	186,500	Fastest conventional zonal rotor.
	--	50,000	2.74×10^7	Edge loading	570	6.4	176,000	Sample may be loaded or unloaded at the center or the edge. MACS option.
MSE B-XIV Ti	B-XIV	47,000	2.42×10^7	Center loading	650	6.7	165,000	Fastest conventional zonal rotor.
MSE B-XIV Al	B-XIV	35,000	1.34×10^7	Center loading	650	6.7	91,000	
Spinco Ti 14	B-XIV	48,000	2.53×10^7	Center loading	665	6.7	172,000	Fastest conventional zonal rotor.
	--	48,000	2.53×10^7	Edge loading	544	6.4	164,000	Sample may be loaded or unloaded at the center or the edge.
Spinco Al 14	B-XIV	35,000	1.34×10^7	Center loading	665	6.7	91,000	
	--	35,000	1.34×10^7	Edge loading	544	6.4	87,000	Sample may be loaded or unloaded at the center or the edge.

Electro-Nucleonics K-5	K-V	35,000	1.34×10^7	Center loading	8390	6.6	90,000	Alternate core configuration of K-type continuous-flow centrifuge (see Table 4).
Electro-Nucleonics RK-5	--	35,000	1.34×10^7	Center loading	4180	6.6	90,000	Alternate core configuration of RK-type continuous-flow centrifuge (see Table 4).

Current zonal rotors (Table 2) are the second- and third-generation descendants from the original rotors and are substantially altered from the first designs. The continuing evolution of zonal rotors is leading not only toward more variety in sizes, shapes, speeds, and special functions, but toward designs that bridge the present distinctions with swinging-bucket rotors. A decade hence we may speak simply of rotors for density-gradient centrifugation.

For the present, zonal rotors are employed mostly for rather large-scale preparative separations of particles.

II. SEDIMENTATION THEORY

In zonal rotors we are concerned almost exclusively with the behavior of cylindrical zones of particles migrating through density gradients. The theory of zonal centrifugation is therefore different--more restricted and more complicated--than that of either differential or analytical sedimentation. We present here a semiquantitative treatment of zonal theory; more rigorous expositions are given elsewhere [3,4].

We shall start with the several elements governing the sedimentation rate of a particle:

$$\frac{dr}{dt} = S^* r \omega^2 \frac{\eta_w}{\eta_m(r)} \frac{\rho_p - \rho_m(r)}{\rho_p - \rho_w} \qquad (1)$$

where $\frac{dr}{dt}$ = the radial movement of the particles, S^* = the sedimentation coefficient of an ideal spherical particle that moves at the same rate as the real particles being studied, r = the distance of the particle zone from the axis of rotation, ω = angular velocity, η = viscosity, ρ = density, and subscripts w, p, and m refer to water, particles, and medium, respectively.

Since the particles are in a gradient, the viscosity η_m and density ρ_m of the medium are not constant, but rather are continuous functions of the radial or volumetric distance of the particle zone from the axis of rotation; they are expressed, therefore, as $\eta_m(r)$, $\rho_m(r)$, etc.

If Eq. 1 is expressed in cgs units, one can compute directly the rate of movement of the particle zone in cm per sec. Sedimentation coefficients are expressed in Svedbergs (= 10^{-13} sec). Equation 1 requires that the angular velocity be expressed in rad/sec. Conversion from rpm is

$$1 \text{ rad/sec} = \frac{\pi}{30} \text{ rpm} \qquad (2)$$

Inspection of Eq. (1) reveals two properties of the particle which affect the sedimentation rate: S^* and ρ_p. Separations based on differences in S^* are said to be <u>rate</u>, <u>rate-zonal</u>, or <u>S-rate</u> separations; those based on differences in ρ_p are called <u>equilibrium-density</u> or <u>isopycnic</u> separations. These two kinds of separations are illustrated in Fig. 3. In rate separation the rates of movement of two kinds of particles are simply proportional to S*. In the case of ideal* spherical particles, the sedimentation coefficient is principally a function of size[†]; but with real particles, S* may vary with the shape and concentration of the particles, the osmotic pressure, the presence of electrolytes, the presence of other particles, etc. Nonetheless, we can say that as a first approximation, rate separations are based on differences in size.

If sedimentation continues the particles will find themselves in a medium of greater and greater density. Ultimately they will arrive in a region where the density of the medium is equal to their own; they then cease to sediment and we say the particles have reached their equilibrium density. In terms of Eq. (1) we note that the sedimentation rate approaches zero as ρ_p approaches ρ_m.

The densities of real particles along with S* are affected by the medium so that ρ_p is not necessarily a constant. We can define the equilibrium density ρ_{eq} as

$$\rho_{eq} \equiv \lim \rho_p \text{ as } \rho_p \rightarrow \rho_m(r) \text{ or as } \frac{dr}{dt} \rightarrow 0.$$

---

*Ideal particles are nondeformable and interact neither with the medium nor with one another.

[†] $S^* = \dfrac{(\rho_p - \rho_w)\alpha^2}{18\eta_w}$ where α is the radius of an ideal sphere.

Isopycnic separations, therefore, are based on differences in equilibrium densities of the particles.

Anderson [2,3] has pointed out that the S* and ρ_{eq} values assigned to subcellular organelles can be plotted on Cartesian coordinates; the particles are then located in "S-ρ space" (Fig. 4). This concept is useful in several ways: e.g., inspection of an S-ρ plot of a mixture of particles can suggest tactics for fractionation or purification. Peroxisomes, for example, can be isolated by a rate separation followed by isopycnic sedimentation of the several fractions obtained. Another representation of this strategy is shown in Fig. 5.

III. GRADIENT SHAPES--RATE SEPARATIONS

A. Resolution

For the first ten years of density-gradient centrifugation, the gradient played a passive role of supporting the particle zones and resisting convective flow. In 1961 Martin and Ames [5] showed that a selected gradient could cause the sedimentation rates of particles--specifically proteins--to be proportional to their sedimentation coefficients. This concept is so fundamental to the further development of density-gradient centrifugation that we should examine its basis.

The sedimentation rates of particles are influenced by opposing forces as they migrate deeper into a gradient [see Eq. (1)]. On the one hand the increasing value of r tends to increase the sedimentation rate, but the simultaneously increasing values of η_m and ρ_m tend to decrease the sedimentation rate. Martin and Ames pointed out that, for a given value of ρ_p, one could hope to balance these two forces so that dr/dt would vary only with S and ω^2. Noll [6] formalized this idea with the term "isokinetic gradients," defined by the relation

$$\frac{dr}{dt} = \text{const} = \frac{r}{\eta_m(r)} [\rho_p - \rho_m(r)] \qquad (3)$$

Isokinetic gradients are fine for swinging-bucket rotors, but of no advantage within the cylindrical coordinates of zonal rotors. The problem here is <u>radial</u> or <u>sectorial dilution</u>, the phenomenon in which constant radial zone width leads to a progressive decrease in the concentration of particles.

At first it seemed as though the advantage of large capacity conferred by cylindrical geometry would be partially cancelled by radial dilution. Schumaker [7] then suggested that gradients could be used to control the shapes of particle zones. Specifically, Schumaker reasoned that one should be able to <u>decrease</u> zone widths if the gradient were sufficiently steep (Fig. 6). In such a gradient the particles in the leading edge of a zone would be decelerated with respect to those in the trailing edge. As a consequence, the zone would tend to collapse on itself.

Schumaker's insight liberates zonal centrifugation from the radial coordinates of classical ultracentrifugation, and permits us to think and work within the volumetric coordinates of the rotor volume. For example, the volumetric analog of an isokinetic gradient would be one in which particles migrate through equal increments of volume with time; these are called <u>equivolumetric gradients</u> [8] (see Fig. 7):

$$\frac{dV}{dt} = 2\pi hr \frac{dr}{dt} = \text{const}$$

$$\frac{r^2}{\eta_m(r)} [\rho_p - \rho_m(r)] = \text{const} \qquad (4)$$

In an equivolumetric gradient the volumetric distance that particles move should be proportional to their sedimentation coefficient and the width of particle zones should remain independent of the distance migrated. These properties are shown for polysomes in Fig. 8 (see also Ref. 9). Berns et al. [10] have obtained exceptionally fine resolution of RNA's in an equivolumetric gradient (Fig. 9).

A compilation of equivolumetric gradients is presented in the Appendix.

Note that the equations for both isokinetic and equivolumetric gradients are derived for particle zones of negligible width. Spragg et al. [10a] computed an <u>isometric</u> gradient from a much more complex model that included finite zone width and the time-dependent diffusion of gradient solutes. They found that their gradient permitted improved resolution of globulins, but they also encountered anomalous zone broadening, a time-dependent phenomenon that detracts from the expected resolution (see also Ref. 11).

We may define resolution in a zonal separation as (see Fig. 8)

$$\Lambda \equiv \frac{\Delta V}{\sigma_1 + \sigma_2} \tag{5}$$

Anything that increases the separation of particle zones ΔV and decreases their width (taken as the standard deviations of Gaussian distribution σ) will improve resolution. In general we find that gradient slopes affect these two quantities in opposite directions, but no general theory on the optimization of resolution has been forthcoming.

An extreme example of gradient-induced zone narrowing is the use of step gradients [12], with which Cline et al. observe evidence of remarkable rate heterogeneity of particles from cell breis. The hydrodynamics of particle movements through step gradients is very complex and gradient capacities are certain to be low, but the general method deserves close examination.

B. Capacity

We speak of the capacity of a gradient as the mass m of particles which can be loaded onto a gradient without having any region become more dense than the underlying gradient. This notion of capacity is valid only instantaneously, since the diffusion of solutes and the sedimentation of particles can and do cause perturbations in particle zones that can make initially stable zones become unstable.

Let us first calculate the instantaneous capacity of different configurations of particle zones (Fig. 10). The density of a solution is increased by particles according to

$$\rho_t = \rho_m + c\overline{V}_p(\rho_p - \rho_m) \tag{6}$$

where ρ_t is the density of the total system t, the medium m, or the particles p; c is the concentration of particles p; and $\overline{V}_p$ is the partial specific volume of the particles (= $1/\rho_p$).

A <u>rectangular</u> sample zone, which is the most common form of sample zones, should remain stable according to the "instantaneous" criterion defined above as long as the density of the particle zone is no greater than ρ_2 at the top of the gradient (Fig. 10). Maximum capacity will occur, therefore, when

$$m \simeq \frac{\rho_p(\rho_2 - \rho_1)}{\rho_p - \rho_1} V_s \tag{7}$$

where V_s is the volume of the particle zone.

Following the same reasoning, a *wedge-shaped* particle zone, also known as an *inverse particle gradient*, has a capacity of [13]

$$m \simeq \frac{1}{2}\left(\frac{\rho_p}{\rho_p - 1}\right)\left(\frac{d\rho}{dV}\right)V_s^{\,2} \tag{8}$$

Gaussian zones [14] have an instantaneous capacity of

$$m \simeq \frac{\sqrt{2\pi\varepsilon\sigma}^{\;2}\;\rho_p(d\rho_m/dV_z)_o}{\rho_p - \rho_m(V_z) - 2\sigma\;(d\rho_m/dV)_o} \tag{9}$$

where ε is the base of the natural logarithm, σ is the standard deviation of the particle distribution, V_s is measured from the center of the particle zone, and subscript o designates the center of that zone.

Among these three zone shapes, the rectangular zone has the largest instantaneous capacity, but it is also most susceptible to a time-dependent instability known as "droplet sedimentation" [1].

It is now generally agreed that droplet sedimentation is caused by the diffusion of solute molecules from the top of the gradient to the particle zone across the infinitely steep concentration gradient. This causes perturbations in the form of locally dense regions of particles-cum-solutes, which bulge down into the gradient, absorb still more solute molecules, and then sink more or less rapidly through the gradient. The theory of droplet sedimentation is not agreed upon [15,16], but Halsall and Schumaker [17a] have determined empirically that a system such as that in Fig. 10 will be stable when the densities of the particle zone and underlying gradient and the diffusivities of particles and solute molecules meet a limiting relation:

$$\frac{\Delta\rho_p}{\Delta\rho_s} = \frac{C_p(1 - \bar{V}_p\,\rho_m'')}{\rho_s' - \rho_s''} < \frac{D_p}{D_s} \tag{10}$$

where C_p is concentration of the particle; $\bar{V}_p$ is partial specific volume of the particle = $1/\rho_p$; ρ_s' is the density of the solute under the sample zone; ρ_s'' is the density of the solute in the sample zone; and D_p and D_s are the diffusion constants of particle and solute.

What this means is that the stability--we should specify static capacity--of a rectangular particle zone will be increased by a high molecular weight of the gradient solute, and decreased by high particle weights.

The instability of rectangular particle zones is clearly traceable to the discontinuity in solute concentration between gradient and particles. Britten and Roberts [18] proposed wedge-shaped sample zones (Fig. 10) in which the solute gradient extends through the sample zone. The static capacity of these zones has not been fully tested, but does appear to approach the expectation of Eq. (8).

Although particle zones are often recovered from zonal rotors in Gaussian distributions, the capacity of such zones has not been probed.

We have spoken of instantaneous and static capacities; there is also a dynamic capacity, which differs from both. Berman [19] suggested that a particle zone that was initially stable could become unstable by gradient-induced zone narrowing. Specifically, Berman asked that we consider a class of hyperbolic gradients:

$$\rho_m(r) = \rho_p - \frac{k}{r^n} \tag{11}$$

where k and n are arbitrary and independent constants.

Berman examined dynamic stability in response to differing values of n. He deduced that for $0 < n \leqslant 1$, the system will always be stable. For $1 < n \leqslant 2$ the system may become unstable as the particles migrate into the gradient. For $2 < n$, the system will become unstable the instant the particles enter the gradient.

Spragg and Rankin [20] and Eikenberry et al. [13] tested Berman's predictions and found them to be largely sustained. Eikenberry et al. in particular showed that up to 2 g of ribosomes could be resolved into subunits in a single hyperbolic gradient of 800 cc in a B-XV rotor. This gradient is compiled in Table 3.

TABLE 3

Gradient Computed for the Separation of Particles of 1.56 gm/cc in a B-XV Zonal Rotor

Volume ml	Radius cm	Density at 5°C g/cm	Sucrose concentration at 5°C % (w/w)	Computed factor for gradient generator[a]
0 to 708	1.9 to 6.0	1.0	0.0	0
750[b]	6.15	1.013[b]	3.4	0.077
800[b]	6.32	1.028[b]	7.1	0.164
808	6.34	1.029	7.4	--
850	6.49	1.042	10.4	0.244
900	6.66	1.055	13.4	0.318
950	6.82	1.067	16.1	0.387
1000	6.98	1.078	18.6	0.451
1050	7.13	1.088	20.9	0.511
1100	7.28	1.098	23.0	0.567
1150	7.42	1.107	24.9	0.619
1200	7.57	1.115	26.7	0.669
1250	7.71	1.124	28.5	0.711
1300	7.85	1.131	30.0	0.763
1350	7.98	1.139	31.5	0.805
1400	8.12	1.146	32.9	0.846
1450	8.25	1.152	34.2	0.886
1500	8.38	1.159	35.5	0.924
1550	8.52	1.165	36.8	0.962

TABLE 3 cont'd.

Volume ml	Radius cm	Density at 5°C g/cm	Sucrose concentration at 5°C % (w/w)	Computed factor for gradient generator[a]
1600	8.66	1.171	38.0	1.000
1650	8.88	1.208	45.0	--

[a]Fraction of program cam height for the Beckman Spinco Model 141 Gradient Pump = volumetric fraction of 38% (w/w) sucrose required to achieve desired density.

[b]These figures computed for the hyperbolic gradient are listed for reference only since in operation this region is normally occupied by the sample zone.

IV. GRADIENT SHAPES--ISOPYCNIC SEPARATIONS

Linear gradients, in which the concentration of solute is proportional to volume, have no special virtues in rate separations. However, they are the usual gradients of choice for isopycnic separations, primarily for the convenience of interpolating the equilibrium density of a particle zone.

A. Resolution

Meselson et al. [21] deduced that isopycnic particle distributions tend to be Gaussian with respect to radius. The standard deviation has been calculated to be

$$\sigma_r = \sqrt{\frac{RT\rho_o}{M^*(d\rho/dr)_o \omega^2 r_o}} \tag{12}$$

where R is the universal gas constant; T is the absolute temperature; M* is the particle weight in daltons; and the subscript o denotes the center of the particle zone.

When such a particle zone is recovered from a cylindrical rotor, it will no longer be Gaussian, but we can obtain an approximate measure of the zone width by transforming σ_r into volumetric units:

$$\sigma_V = 2\pi rh \, \sigma_r$$

Substituting this relation into Eq. (12) and assuming the rotor to be a simple cylinder, we obtain

$$\sigma V = \sqrt{\frac{2\pi h \; RT\rho_o}{M^*(d\rho/dV)_o \omega^2}} \tag{13}$$

It follows that for a given rotor, σ_V is decreased by a decrease in both the slope $d\rho/dV$ and the speed ω.

Since resolution [Eq. (5)] is determined by the separation between zones as well as by zone width, we can ask how resolution would be affected by the gradient slope. Since separation between zones varies inversely with the slope but zone width varies inversely with the square of the slope, we must conclude that resolution is better with shallower gradients.

Equation (13) tells us that resolution is independent of the position of the zone in the rotor r, although we can expect that equilibrium will be approached more rapidly at the higher centrifugal fields existing near the edge of the rotor.

The zone widths of Eq. (13) are those predicted for homogeneous particles. Larger σ values would indicate heterogeneity with respect to particle size M* or density ρ_{eq}. Expected zone widths of different-size particles under arbitrary conditions of centrifugation are plotted in Fig. 11.

B. Capacity

There is no inherent limit to gradient capacity in isopycnic separations. I have seen zones so concentrated as to be semisolid.

V. GRADIENT MATERIALS

The properties of the ideal gradient material are stated simply:

- freely soluble in water
- very dense
- nonviscous
- negligible osmotic pressure
- physiologically and chemically inactive
- transparent in visible and ultraviolet light
- cheap

A. Sucrose

Sucrose, for all its faults, comes as close to ideal as any presently available gradient material (Table 4).

TABLE 4

Physical Properties of Some Gradient Materials (Compiled from Various Sources)

	Stock solution (stable at 4°)			
Substance	Concentration	Density in g/cc (°C)	Viscosity in centipoise	Refractive index at 20°C
CsCl	60% w/w	1.7900 (20°)	--[a]	1.4074
D_2O	100%	1.105 (20°)[b]	--	1.33844
Ficoll	46.5% w/w	1.1629 (4°)	1020	1.3764[c]
Glycerol	100%	1.2609 (20°)	1490 (20°)	1.4729
Silica sols[d]	40.1% w/w	1.295 (25°)	27 (25°)	--
Sorbitol	60% w/w	1.2584 (4°)	102.9 (4°)	1.4402
Sucrose	65% w/w	1.32600 (4°)	56.5 (20°)	1.4532
Various synthetic organics which are highly soluble in water[e]				

[a]Determined by Kaempfer and Meselson [22] at 0°C as a function of density.

[b]The density of CsCl solutions at 25°C has been computed as a function of the refractive index by Ifft et al. [23]; $\rho = 10.8601$ (RI) - 13,4974.

[c]A 30% w/v solution.

[d]Available under the trade name of Ludox (Du Pont de Nemours Chemical Co.).

[e]Parish et al. [24].

It is insufficiently dense for isopycnic separations of particles whose ρ_{eq}'s are greater than 1.3 and it probably should not be used other than as an osmoticum with the more delicate membrane-bound organelles, but for most purposes it serves very well. For experimenters with direct access to computers, it is probably simplest to store the empirical constants, as determined by Barber [25], relating sucrose concentration to viscosity and density.

1. In the range of 0° to 30°C, the density of sucrose may be related to temperature T and weight fraction Y by the relation

$$\rho_{T,m} = (B_1 + B_2T + B_3T^2) + (B_4 + B_5T + B_6T^2)Y + (B_7 + B_8T + B_9T^2)Y^2$$

where

$\rho_{T,m}$ = density of a sucrose solution

T = temperature, °C

Y = weight fraction sucrose

and the B_1's are constants.

Constant	Value[a]
B_1	1.0003698
B_2	3.9680504×10^{-5}
B_3	$-5.8513271 \times 10^{-6}$
B_4	0.38982371

Constant	Value[a]
B_5	$-1.0578919 \times 10^{-3}$
B_6	1.2392833×10^{-5}
B_7	0.17097594
B_8	4.7530081×10^{-4}
B_9	$-8.9239737 \times 10^{-6}$

[a]Values are given to 8 figures for machine calculations; use of the first 5 figures would be sufficient for hand calculations.

2. The viscosity of sucrose η between 0° and 80°C can be expressed as a fraction of temperature T and the mole fraction y:

$$\log \eta_{T,m} = A + \frac{B}{T + C}$$

The mole fraction y is related to the weight fraction Y and the molecular weights of sucrose S and water W by the relation

$$y = \frac{Y/S}{Y/S + (1-Y)/W}$$

The constants A and B are calculated from y by the relations

$$A = D_o + D_1 y + D_2 y^2 + D_3 y^3{}_7 + \ldots + D_\eta y^\eta$$

$$B = E_o + D_1 y + E_2 y^2 + E_3 y^3{}_7 + \ldots + E_\eta y^\eta$$

Coefficients[a]	Range of equation, wt %	
	0 to 48	48 to 75
D_o	-1.5018327	-1.0803314
D_1	9.4112153	-2.0003484×10^1
D_2	-1.1435741×10^3	4.6066898×10^2
D_3	1.0504137×10^5	-5.9517023×10^3
D_4	-4.6927102×10^6	3.5627216×10^4
D_5	1.0323349×10^8	-7.8542145×10^4
D_6	-1.1028981×10^9	
D_7	4.5921911×10^9	
E_o	2.1169907×10^2	1.3975568×10^2
E_1	1.6077073×10^3	6.6747329×10^3
E_2	1.6911611×10^5	-7.8716105×10^4
E_3	-1.4184371×10^7	9.0967578×10^5
E_4	6.0654775×10^8	-5.5380830×10^6
E_5	$-1.2985834 \times 10^{10}$	1.2451219×10^7
E_6	1.3532907×10^{11}	
E_7	$-5.4970416 \times 10^{11}$	

[a]Coefficient subscript indicates the exponent of the composition by which the coefficient is to be multiplied.

The value of C is related to the weight fraction y plus three additional constants:

$$C = G_1 - G_2 \left[1 + \left(\frac{y}{G_3}\right)^2\right]^{\frac{1}{2}}$$

$G_1 = 146.06635$, $G_2 = 25.251728$, $G_3 = 0.070674842$

Sucrose should be purified before use by passing 65% w/w solutions through a mixed-bed ion-exchange column 1 cm in diameter and about 50 cm long. The output can be monitored with a conductivity meter. Commercial sucrose may be quite acceptable as a starting material and substantially less expensive than the reagent or analytical grade material.

B. Sorbitol

Sorbitol has some advantages over sucrose for membrane-bound organelles. Although the increase in osmotic pressure per increment of density is somewhat greater than for sucrose, the ultrastructure of mitochondria [26] and chloroplasts [27] is better preserved. I might note in passing that mannitol was preferred to sucrose by an older generation of physiologists on the grounds that it permeated cell membranes less rapidly. Sorbitol presumably has similar properties but is much more soluble than mannitol.

A compilation of some properties of sorbitol is presented in Table 5.

TABLE 5

Properties of Sorbitol[a]

Temp = 4°C [Sorbitol] (% w/w)	Density (g/cc)	Viscosity (centipoise)	Refractive index (4°C)	Refractive index (20°C)
1	1.0022	1.8	1.3350	1.3358
5	1.0170	2.0	1.3405	1.3420
10	1.0358	2.4	1.3482	1.3495
15	1.0546	--	1.3560	1.3577
20	1.0752	3.6	1.3640	1.3656
25	1.0964	--	1.3725	1.3741
30	1.1160	6.0	1.3812	1.3834
35	1.1388	7.3	1.3897	1.3920
40	1.1610	10.7	1.3993	1.4011
45	1.1850	15.7	1.4087	1.4060
50	1.2084	25.7	1.4195	1.4210
55	1.2330	46.2	1.4309	1.4312
60	1.2584	102.9	1.4402	1.4425

[a]Taken from Ref. 28.

C. Ficoll

Ficoll* is a synthetic polymer of sucrose and, except for its viscosity and expense, is nearly ideal for membrane-bound organelles. Gorcznski et al. [29] separated mouse spleen cells on Ficoll gradients and showed (1) that B cells could be resolved

*Tradename of Pharmacia Fine Chemicals.

from T cells and (2) that heterogeneity previously found in sedimentation through albumin gradients was attributable to the coexistence of an osmotic gradient in albumin.

We found Ficoll to be the only gradient material among several that we tried that permitted the isolation of intact chloroplasts from Euglena [30,31] (Fig. 12).

Some properties of Ficoll are compiled in Table 6.

TABLE 6

Density, Refractive Index, and Viscosity of Ficoll[a]

wt %	Density (g/ml)				Refractive index (20°C)
	2°C	4°C	10°C	20°C	
0.68	1.000	1.003	1.002	0.998	1.3350
2.09	1.004	1.009	1.006	1.002	1.3360
4.92	1.012	1.019	1.015	1.011	1.3400
8.23	1.021	1.029	1.024	1.022	1.3433
11.75	1.030	1.039	1.033	1.032	1.3484
18.74	1.048	1.059	1.052	1.051	1.3555
25.06	1.064	1.075	1.065	1.067	1.3622
31.22	1.080	1.091	1.083	1.082	1.3688
38.69	1.090	1.107	1.100	1.099	1.3764
48.09	1.113	1.125	1.121	1.227	1.3837

TABLE 6 cont'd.

wt %	Viscosity (centipoise) 2°C	4°C	10°C	20°C
0.68	13	14	12	10
2.09	16	16	14	13
4.92	16	16	15	16
8.23	20	20	19	17
11.75	22	20	20	20
18.74	39	34	30	24
25.06	60	51	49	35
31.22	91	90	79	58
38.69	159	151	122	90
48.09	295	273	212	157

[a]Taken from Ref. 32.

D. Deuterium Oxide

Deuterium oxide has the advantage of being virtually indistinguishable from water from a physiological standpoint, but its extreme expense and relatively low density have prohibited any widespread use in zonal centrifugation.

E. Cesium Compounds

The high cost of cesium chloride or cesium sulfate has also restricted their use in zonal rotor. Most multimolecules and more-complex particles are at least partially degraded by sedimentation into high concentrations of their alkali salts, but the extreme densities obtainable make them essential for some separations.

Cesium salts have been used in edge-loading zonal rotors (see Section VI, Techniques) in thin annular gradients into which particles can be collected by isopycnic sedimentation. The gradient is then recovered by displacement with water.

Empirical equations relating concentration, density, and refractive index of cesium salts are presented in Table 7.

TABLE 7

Coefficients for Calculation of Density, from Refractive Indexes of Solutions at 25°C[a]
(Coefficients are presented for the relation $\rho = a\eta_D - b$ where η_D is the refractive index.
All measurements are at 25°.)

Solute	Coefficients a	b	Density range
Cs_2SO_4	12.1200	15.1662	1.15-1.40
	13.6986	17.3233	1.40-1.70
CsBr	9.9667	12.2876	1.25-1.35
CsCl	10.8601	13.4974	1.25-1.90
Cs acetate	10.7527	13.4247	1.80-2.05
Cs formate	13.7363	17.4286	1.72-1.82
KBr	6.4786	7.6431	1.10-1.35
RbBr	9.1750	11.2410	1.15-1.65

[a]Taken from Ref. 33.

Ludlum and Warner [34] give a different formula for density $\rho_{25°}$ of Cs_2SO_4:

$$\rho_{25°} = 0.9954 + 11.1066(\eta_D - \eta_{D,o}) + 26.4460(\eta_D - \eta_{D,o})^2$$

where η_D is the refractive index of the solution and $\eta_{D,o}$ that of water.

F. Silica Sols

Silica sols marketed under the tradename of Ludox are of increasing interest for density-gradient centrifugation. The silica particles are on the order of 100 Å in diameter; despite their highly charged surfaces, the osmotic potential of Ludox solutions is negligible. They are quite transparent, are somewhat viscous, and when combined with other polymers can be made physiologically inactive [35,36]. To their substantial disadvantage, they gel rapidly at pH below 7 and in the presence of moderate concentrations of cations.

Pertoft [37] has isolated Herpes virus by isopycnic sedimentation in Ludox and we have separated intact from stripped spinach chloroplasts in Ludox by continuous-flow centrifugation with banding (Fig. 13) (see also Ref. 38).

We should note that the equilibrium densities of membrane-bound particles, including whole cells, are less--often dramatically so--in a non- or isosmotic gradient of a polymer than in, say, sucrose. The reason, of course, is that water is drawn out of the particles and they become denser in gradients of high osmotic potential.

A table of properties of the several formulations of Ludox is presented in Table 8.

G. Other Materials

Other rarely used gradient materials are glycerol, Renograffin*, sodium iodothalamate, albumin, polyethylene-glycols, and polyvinylpyrrolidone.

These salts may be added to gradients to increase their densities: potassium citrate or potassium tartrate, and sodium bromide [39].

Nonaqueous gradients may also be prepared [40], but these have not been reported for zonal rotors.

VI. TECHNIQUES

A. Gradient Generators and Pumps

The ideal gradient maker should be completely programable†, capable of pumping up to 1 liter of quite viscous solutions in 30 min, and easy to clean and sterilize.

---

*Urograffin, a radiologically contrasting agent, very dense, but opaque in the UV: methylglutamine salt of N,N'-diacetyl-3,5-diamino-2,4,6-triiodobenzoic acid.

†Capable of generating a gradient of any desired shape.

TABLE 8

Properties of Ludox Colloidal Silica[a]

Formulation	HS		LS	AS	AM	SM
	"40%"	"30%"[b]				
% Silica as SiO_2	40.1	30.1	30.1	30.1	30	15
% Na_2O (titratable alkali)	0.43	0.32	0.10	0.25[c]	0.13	0.10
Chloride as % NaCl	0.03	0.01	0.001	0.002	0.007	0.001
Sulfate as % Na_2SO_4	0.05	0.04	0.010	0.007	0.006	0.003
Viscosity at 25°C, cps	27	5.1	10.4	16	10	4
pH at 25°C	9.6	9.8	8.3	9.4	9.1	8.5
Approximate particle diameter, millimicrons	13-14	13-14	15-16	13-14	13-14	7-8
Surface area, M^2/g (B.E.T. method)[d]	220-235	220-235	195-215	220-235	220-235	350-400
Specific gravity @ 25°C	1.295	1.211	1.209	1.206	1.209	1.093

KEY:
HS--High Sodium stabilization level
LS--Low Sodium stabilization level (by contrast to HS grade)
SM--Seven Millimicron (refers to particle size
AS--Ammonia Stabilized (ammonium hydroxide used as alkali for stabilization)

[a]Information courtesy of E. I. du Pont de Nemours & Co. (Inc.), Industrial and Biochemicals Department, Wilmington, Delaware 19898.

[b]Freeze stabilized form available.

[c]% ammonia.

[d]Nitrogen adsorption on dry solids; see P. H. Emmett, Symposium on "New Methods for Particle Size Determination" p. 95, Pub. by ASTM March 4, 1941.

There are three commercial devices that are fully programable (Table 9), but none have yet been proven fully successful on all counts.

Several commercial devices are available that will generate either linear or exponential gradients. Linear gradients are suitable for isopycnic separations and are often used but are not optimal for rate separation; exponential gradients with some fuss can often be made to approximate hyperbolic and equivolumetric gradients.

Anderson and Ruttenberg [41] described an elegantly simple and inexpensive exponential generator suitable for zonal rotors that may be fabricated from laboratory supplies (Fig. 14) or is commercially available.

Simple two-cylinder generators can be prepared from plexiglass or glass columns (Fig. 15), but vigorous stirring in the mixing chamber is essential; a "Vibromixer" (Chemag, A. G., Mannedorf/ZH, Switzerland) is suitable. I should caution against the coefficient of difficulty that arises when one tries to scale up without modification the small gradient devices suitable for swinging-bucket rotors to the liter and more of gradient volume required for zonal rotors. If one can avoid airlocks while the gradient is introduced into the zonal rotor, the solution will actually be drawn into the spinning rotor, but substantial backpressures are commonly encountered. A positive-displacement pump that is capable of working against at least 2 atm is usually need peristaltic pumps usually fail with viscous gradients.

Characteristics of Commercially Available Gradient Generators

Manufacturer and designation	Kinds of gradients delivered			Maximum volume (cc)	Maximum pumping rate (cc/min)	Suitability with viscous materials	Ease of sterilization
	Linear	Exponential	Fully programable				
IEC 3651	+	+		2000	70	+++	++
IEC 3660		+		a	ca. 100	++	+++
ISCO Dialagrad 380			b		60	++	+
ISCO Dialagrad 382			b		10	++	+
ISCO 570	+	+		80	16	+++	+
ISOLAB Gradipore			+	1500	c	-	-
LKB GM-1	+[e]			500	c	+	
Ultragrad			+		d	d	+++
MSE fixed profile	+[e]			2000	c	+	+++
MSE automatic variable			+				+++
Sorvall	+	+			c		
Spinco 141			+	6000	66	-	-

[a]Depends on selection of mixing chamber.

[b]Gradient determined by smoothed fit to 11 specified points along gradient.

[c]Depends on gravity or choice of auxiliary pump.

[d]Pumps suitable for sucrose have not been fully described.

Other considerations in choosing a pump concern the recovery of the gradient. Normally one displaces the gradient from the rotor by pumping a denser solution to the edge or a lighter solution (water) to the center. The pump for this should be pulseless and, since one may want to collect fractions by time, should be of constant controllable speed.

B. Choice of Rotor

The list of current, batch-type zonal rotors in Table 2 is bewildering at first glance, but the actual choices among rotors are more limited. Since rotors are not yet (unfortunately) interchangeable, we are usually limited to the available centrifuge drive units, which are dictated in turn by local preferences, servicing, etc.

In our group, research workers chose a B-XIV-type of rotor for eight out of ten runs. Even though the high speed (up to 50,000 rpm) may not be required, the volume of the rotor (about 650 cc) is convenient for most separations, and the rotor is physically easy to handle. The edge-loading mode is also an attractive option. The titanium B-XIV's are generally suitable for rate separations of particles as small as 5 S. If I were restricted to a single rotor, I should unhesitatingly select a B-XIV type.

For larger gradient volumes (ca. 1.5 liters), we turn to the B-XV and B-XXIX. These two types of rotors have virtually the same geometry, except that the distal surface of the B-XXIX is shaped for edge-loading and unloading. The titanium B-XV's and B-XXIX's are generally suitable for rate separations of particles as small as 15 S.

The type JCF-Z rotor appears to be roughly comparable to the B-XV, but little performance data have been published.

The K- and RK-series were designed as large-scale continuous-flow-with-banding rotors, in which a particle suspension is passed along an annular gradient. The K-V and RK-5 rotors are for conventional (i.e., batch-type) rate or isopycnic separations, and thus provide a welcome versatility to the K- and RK-drive units. However, their high costs and relatively large volumes put them in a special category.

The transparent A-XII, Z-15, and HS rotors permit one to observe the course of sedimentation either visually or photometrically (Fig. 16). They thus form a special class of analytical-preparative zonal rotors. Their speeds are relatively low and their costs high, but they are most useful in developing optimum conditions for particle separations and in estimating sedimentation rates [42]. These rotors are ideal for resolving whole cells [43] and membrane-bound organelles, but may be used for particles as small as 1000 S.

The special case of the Sorvall SZ-14 is discussed below under subsection D.

C. Operation of Zonal Rotors

The several steps involved in carrying out a conventional zonal separation are illustrated in Fig. 17. (a,b) The gradient is pumped in through the edge line with the fluid seal in place; the rotor is kept at a "parking speed" (500 to 3000 rpm) chosen to be safely above any low-speed "criticals" or resonant frequencies. In order to fill the rotor a quantity of dense solution, called underlay, is pumped in under the gradient. (c) The sample is backed in through the centerline, displacing part of the underlay. (d) An overlay then follows the sample. This component plays an important role in zonal centrifugation, although it has no counterpart in the familiar practices of density-gradient centrifugation in tubes: first, the volume of the overlay determines the radial position of the sample zone; second, by a kind of reverse radial dilution, the further the sample zone is displaced the narrower it becomes; and third, the presence of a gradient overlay improves resolution by minimizing the diffusion of solvent into the sample zone [44]. (e) The fluid seal is replaced with a vacuum cap and accelerated to the desired operating speed. (f) At the end of the run, the rotor is returned to its "parking speed," the fluid seal is reattached, and the gradient is displaced from the center (center unloading) by the pumping of underlay to the edge or (as shown in the figure) the gradient is collected from the edge (edge-unloading) by the pumping of water to the center. Although edge unloading has several handsome advantages, not the least of which is having at the end of a run a rotor filled with water rather than 60% w/w sucrose, it does involve local density inversions; so that resolution suffers somewhat. It is also essential to purge the rotor of airlocks before attempting to displace the gradient against a natural pressure gradient. (g) The displaced gradient may be led through one or more flow monitors (absorbancy at one or more wavelengths, refractive index, light scattering, radioactivity, etc.) and collected in fractions.

D. Reorienting Gradient Centrifugation

In density-gradient centrifugation in either swinging buckets or zonal rotors, the isodense layers of the gradient maintain a constant orientation with respect to the container, but there are certain advantages in introducing a gradient into a rotor at rest and allowing the gradient to reorient during acceleration and deceleration [45,46] (Fig. 18). This general method is now the method of choice for the isopycnic separation of small quantities of nucleic acids in angle rotors [47] and has been used for DNA in zonal rotors [48].

The principal advantage of reorienting gradient centrifugation is the elimination of the fluid shear that can occur as particle suspensions traverse the fluid seals of conventional zonal rotors. Native DNA [17] and chloroplasts [49], for example, are partially degraded in crossing standard zonal seals, but can be recovered intact if the fluid lines are widened. The method will presumably be generally advantageous for delicate structures.

The problem with all reorienting gradient techniques is to ensure that the gradient itself is not altered by either shear or Coriolis forces (see Ref. 50); exquisitely smooth acceleration and deceleration are required between 0 and 800 rpm. The problem is more acute in a large, bowl-shaped zonal rotor than in a small, angle rotor or in the thin, annular gradient volumes of K-type continuous-flow rotors. In general, an electronically controlled "ramp generator" is required to program smooth acceleration and deceleration.

Sheeler and Wells [51] have described a reorienting gradient rotor (available commercially from Sorvall as the SZ-14) that seeks to exploit the advantages in this technique. They emphasize, I think wrongly, the convenience and economy achieved by the elimination of fluid seals. Although Sheeler and Wells present some data on the fractionation of cell breis, we do not know how closely reorienting gradient rotors can match the resolution obtainable with dynamically loaded and unloaded rotors.

E. Continuous-Flow Harvesting

Rotors for the collection of particles from large volumes of suspensions have been widely employed for many years. The prototype of continuous-flow harvesting is the Sharples centrifuge in which the suspension is led through a hollow spinning cylinder and the particles are pelleted against the wall. The K-type zonal rotor (Fig. 19) employs the same principle, except that the particles are sedimented into an annular density gradient at the edge of the rotor. The particles are first trapped within the gradient and ultimately they move to their equilibrium density. Therefore, the particles are not only concentrated many fold from the original suspension, but recovered according to their ρ_{eq}.

A small number of such zonal rotors is commercially available, often as optional configurations of batch-type rotors (Table 10). Except for the production-scale K rotors, gradient volumes tend to be quite modest (several hundred cc); this is no disadvantage, since the capacities of gradients in isopycnic separations are enormous. Furthermore, one is often obliged to employ expensive gradient materials in these separations.

The critical property of a continuous-flow system is clean-out. This is the amount of particles collected as a fraction of the amount in original suspension. One usually measures clean-out as

$$f = \frac{\text{initial concentration - concentration in effluent}}{\text{initial concentration}}$$

TABLE 10

Characteristics of Continuous-Flow Zonal Rotors

Manufacturer and designation	Gradient volume (cc)	Maximum speed		Radius at edge	g_{max}	Flow-path length (cm)	Drive unit
		rpm	ω^2				
IEC CF-6	515	6,000	3.95×10^5	14.7	5,930	92.3	PR-2, PR-6, PR-6000
Spinco JCF-Z	400	20,000	4.39×10^6	8.9	40,000	8.9	J-21
Spinco CF-32 Ti	305	32,000	1.12×10^7	8.9	101,300	7.3	L-series
Electro-Nucleonics							
K-3	3200	35,000	1.34×10^7	6.6	90,000	76.2	Interchangeable cores in K-series bowl and drive unit.
K-10	8000	35,000	1.34×10^7	6.6	90,000	76.2	
K-11	380	35,000	1.34×10^7	6.6	90,000	76.2	
Electro-Nucleonics							
RK-3	1600	35,000	1.34×10^7	6.6	90,000	38.1	Interchangeable cores in RK bowl and drive unit.
RK-10	3980	35,000	1.34×10^7	6.6	90,000	38.1	
RK-11	190	35,000	1.34×10^7	6.6	90,000	38.1	
Electro-Nucleonics							
J-1	780	55,000	3.32×10^7	4.4	150,000	38.4	RK

Cleanout will obviously be a function of the sedimentation coefficient of the particle, the radial width of the flowing stream, and the residence time of the suspension in the rotor. This latter quantity is a function of the flow rate, the volume of the flowing stream within the rotor, and the geometry of this volume. Sartory [16] has deduced an equation relating these quantities to cleanout. Peradi and Anderson [52] have determined that a thin flow volume improves cleanout. The important take-home lesson for us is that, because of their short sedimentation distance, cleanout in these continuous-flow zonal systems is an order of magnitude better than in conventional continuous-flow rotors that rely on pelleting.

F. MACS Systems

Conventional zonal rotors have two channels, an edge and a center line, that connect with two channels in the fluid seal. In the multiple alternate channel selection (MACS) system, the two channels in the fluid seal connect with any two of a number of channels in the rotor. This is illustrated in Fig. 20.

Three applications of the MACS system come to mind: (1) a third channel for the insertion of a sample zone at a specified radius, as in Fig. 20; (2) the use of completely independent sets of channels for the gradient and for the flow stream in a continuous-flow-with-banding rotor [49,49a]; and (3) a third channel for semibatch harvesting.

APPENDIX. EQUIVOLUMETRIC GRADIENTS

A. In Table A-1, normalized equivolumetric gradients of aqueous sucrose at 4°C for the B-XIV rotor are expressed as the fractional volumetric mixing ratio which is given as a function of volume. The column for $\rho_p = 1.20$ g/cm^3 is an excellent choice for use as a "universal" equivolumetric gradient generation program. In order to complete the filling of the rotor and to properly position the gradient in the rotor at the bottom of the core pyramid, an overlay of about 33 cm^3 is required. The symbols V_c and c_1 are the constants for the exponential approximations to each of these equivolumetric gradients.

TABLE A-1

Volume, cm^3	Volumetric mixing fraction when $\rho_p = 1.10$ g/cm^3	$\rho_p = 1.15$ g/cm^3	$\rho_p = 1.20$ g/cm^3	$\rho_p = 1.40$ g/cm^3
V_c, cm^2	150	163	176	184
c_1	0.98	0.98	0.99	1.00
0	0.0	0.0	0.0	0.0
25	0.191	0.177	0.170	0.167
50	0.325	0.303	0.291	0.286
75	0.426	0.398	0.385	0.377
100	0.506	0.475	0.460	0.451
125	0.570	0.538	0.522	0.512
150	0.623	0.591	0.575	0.565
175	0.669	0.637	0.621	0.610

TABLE A-1 cont'd.

Volume, cm^3	Volumetric mixing fraction when $\rho_p = 1.10$ g/cm^3	$\rho_p = 1.15$ g/cm^3	$\rho_p = 1.20$ g/cm^3	$\rho_p = 1.40$ g/cm^3
200	0.708	0.677	0.661	0.651
225	0.742	0.713	0.697	0.686
250	0.772	0.744	0.729	0.719
275	0.799	0.772	0.758	0.748
300	0.823	0.798	0.784	0.775
325	0.844	0.821	0.809	0.800
350	0.864	0.843	0.831	0.823
375	0.881	0.863	0.852	0.844
400	0.898	0.881	0.871	0.864
425	0.913	0.898	0.889	0.883
450	0.926	0.913	0.906	0.901
475	0.939	0.928	0.922	0.917
500	0.951	0.942	0.936	0.933
525	0.962	0.955	0.951	0.948
550	0.973	0.967	0.964	0.962
575	0.982	0.979	0.976	0.975
600	0.991	0.990	0.988	0.988
625	1.000	1.000	1.000	1.000

B. Table A-2 is similar to Table A-1, but it is for the B-XV rotor. In order to complete the filling of the rotor, an overlay of about 48 cm^3 is required.

TABLE A-2

Volume cm^3	Volumetric mixing fraction when ρ_p = 1.10 g/cm^3	ρ_p = 1.15 g/cm^3	ρ_p = 1.20 g/cm^3	ρ_p = 140 g/cm^3
V_c, cm^3	276	318	348	368
c_1	0.96	0.96	0.96	0.97
0	0.0	0.0	0.0	0.0
50	0.220	0.198	0.188	0.182
100	0.360	0.328	0.311	0.301
150	0.460	0.422	0.402	0.389
200	0.536	0.495	0.472	0.458
250	0.596	0.553	0.530	0.514
300	0.645	0.602	0.578	0.562
350	0.685	0.643	0.618	0.602
400	0.719	0.678	0.654	0.638
450	0.749	0.709	0.685	0.669
500	0.774	0.736	0.713	0.697
550	0.797	0.760	0.738	0.723
600	0.817	0.782	0.761	0.746
650	0.835	0.802	0.782	0.767
700	0.851	0.820	0.800	0.787
750	0.866	0.837	0.818	0.805
800	0.879	0.852	0.834	0.822
850	0.891	0.866	0.849	0.837
900	0.902	0.879	0.863	0.852
950	0.912	0.891	0.877	0.866

TABLE A-2 cont'd.

Volume cm^3	Volumetric mixing fraction when ρ_p = 1.10 g/cm^3	ρ_p = 1.15 g/cm^3	ρ_p = 1.20 g/cm^3	ρ_p = 1.40 g/cm^3
1000	0.922	0.902	0.889	0.879
1050	0.931	0.913	0.901	0.892
1100	0.939	0.923	0.912	0.903
1150	0.946	0.932	0.922	0.915
1200	0.953	0.941	0.932	0.925
1250	0.960	0.949	0.941	0.935
1300	0.966	0.957	0.950	0.945
1350	0.972	0.964	0.959	0.954
1400	0.978	0.971	0.967	0.963
1450	0.983	0.978	0.975	0.972
1500	0.988	0.985	0.982	0.980
1550	0.993	0.991	0.989	0.988
1600	0.997	0.997	0.996	0.995
1650	1.002	1.002	1.002	1.003

C. The best-fit exponential approximations to normalized equivolumetric gradients of sucrose for the B-XIV and B-XV rotors is given in Fig. A-1.

Computations are by E. F. Eikenberry (57).

ADDITIONAL GENERAL REFERENCES

Cline and Ryel [53], Anderson [54], and Price [55] have reviewed zonal centrifugation in detail. Density gradient centrifugation has been reviewed relatively recently by Moore [56] and Schumaker [58] and is the subject of a forthcoming monograph by Price [4].

In addition a number of useful compilations of zonal methods have recently appeared [59, 60, 61, 62].

REFERENCES

1. N. G. Anderson, in Physical Techniques in Biological Research (G. Oster and A. W. Pollister, eds.), Vol. III, Cells and Tissues, Academic Press, New York, 1956, pp. 299-352.

2. N. G. Anderson, Science, 154, 103-112 (1966).

3. N. G. Anderson (ed.), Zonal Centrifugation, National Cancer Institute Monograph #21, 1966, 526 pp.

4. C. A. Price, Centrifugation in Density Gradients, Academic Press, New York, 1973.

5. R. G. Martin and B. N. Ames, J. Biol. Chem., 236, 1372-1379 (1961).

6. H. Noll, Nature, 215, 360-363 (1967).

7. V. N. Schumaker, Separation Sci., 1, 409-411 (1966).

8. M. S. Pollack and C. A. Price, Anal. Biochem., 42, 38-47, (1971).

9. C. A. Price and T. S. Hsu, Vierteljahrschr. Naturforsch. Ges. Zuerich, 116, 367-375 (1971).

10. A. J. M. Berns, R. A. de Abreu, M. van Kraaikamp, E. L. Benedetti, and H. Bloemendal, FEBS Letter, 18, 159-163 (1971).

10a. S. P. Spragg, R. S. Morrod, and C. T. Rankin, Jr., Separation Sci., 4, 467-481 (1969).

11. S. P. Spragg, in Particle Separation from Plant Materials (C. A. Price, ed.), No. CONF-700119, Oak Ridge National Laboratories Microsymposium, National Technical Information Service, Springfield, Va., 1970.

12. G. B. Cline, R. B. Ryel, and M. K. Dagg, in Particle Separation from Plant Materials (C. A. Price, ed.), Oak Ridge National Laboratories Microsymposium, National Technical Information Service, Springfield, Va., 1970, CONF-70019.

13. E. F. Eikenberry, T. A. Bickle, R. R. Traut, and C. A. Price, Eur. J. Biochem., 12, 113-116 (1970).

14. J. Vinograd and R. Bruner, Biopolymers, 4, 131-156 (1966).

15. H. Svensson, L. Hagadahl, and K. D. Lerner, Sci. Tools, 4, 37-41 (1957).

16. W. K. Sartory, Biopolymers, 7, 251-263 (1969).

16a. W. K. Sartory, Separation Sci., 5, 137-143 (1970).

17. H. B. Halsall and V. N. Schumaker, Nature, 221, 772-774 (1969).

17a. H. B. Halsall and V. N. Schumaker, Biochem. Biophys. Res. Comm., 43, 601-606 (1971).

18. R. J. Britten and R. B. Roberts, Science, 131, 32-33 (1960).

19. A. S. Berman, in The Development of Zonal Centrifuges (N. G. Anderson, ed.), National Cancer Institute Monograph #21, 1966, pp. 41-76.

20. S. P. Spragg and C. T. Rankin, Jr., Biochim. Biophys. Acta, 141, 164-173 (1967).

21. M. Meselson, F. W. Stahl, and J. Vinograd, Proc. Natl. Acad. Sci. U.S., 43, 581-588 (1957).

22. R. Kaempfer and M. Meselson, Methods in Enzymol., 20, 520-528 (1971).

23. J. B. Ifft, D. H. Voet, and J. Vinograd, J. Phys. Chem., 65, 1138-1145 (1961).

24. J. H. Parish, J. R. B. Hastings, and K. S. Kirby, Biochem. J., 99, 19 P (1966).

25. E. J. Barber, in The Development of Zonal Centrifuges (N. G. Anderson, ed.), National Cancer Institute Monograph #21, 1966, pp. 219-239.

26. C. J. Avers, A. Szabo, and C. A. Price, J. Bacteriol., 100, 1044-1048 (1969).

27. D. H. Brown, personal communication, 1970.

28. C. Suerth, unpublished results, 1968.

29. R. M. Gorcznski, R. G. Miller, and R. A. Phillips, Immunology, 19, 817-829 (1970).

30. A. Vasconcelos, M. Pollack, L. R. Mendiola, H. P. Hoffmann, D. H. Brown, and C. A. Price, Plant Physiol., 47, 217-221 (1971).

31. C. A. Price and M. S. Pollack, Vierteljahrsch. Naturforsch. Ges. Zuerich, 116, 409-413 (1971).

32. H. W. Hsu, L. R. Bell, and D. B. Sims, Anal. Biochem. (in press).

33. J. Vinograd and Hearst (1962).

34. Ludlum and Warner (1965).

35. G. M. Mateyko and M. J. Kopac, Ann. N. Y. Acad. Sci., 105, 219-285 (1963).

36. H. Pertoft, O. Back, and K. Lindahl-Kiessling, Exptl. Cell Res., 50, 355-368 (1968).

37. H. Pertoft, Anal. Biochem., 38, 506-516 (1970).

38. J. W. Lyttleton, Anal. Biochem., 38, 277-281 (1970).

39. H. G. Wilcox and M. Heimberg, Biochim. Biophys. Acta, 152, 424-426 (1968).

40. M. Behrens, in Handb. biol. Arbeitsmethoden (E. Abderhalden, ed.), 5: (10-II), pp. 1363-1392.

41. N. G. Anderson and E. Rutenberg, Anal. Biochem., 21, 259-265 (1967).

42. C. A. Price and A. P. Hirvonen, Biochem. Biophys. Acta, 148, 531-538 (1967).

43. J. Sebastian, B. L. A. Carter, and H. O. Halvorson, J. Bacteriol., 108, 1045-1050 (1971).

44. C. A. Price and A. Kovacs, Anal. Biochem., 28, 460-468 (1969).

45. N. G. Anderson, C. A. Price, W. D. Fisher, R. F. Canning, and C. L. Berger, Anal. Biochem., 7, 1-9 (1964).

46. L. H. Elrod, L. C. Patrick, and N. G. Anderson, Anal. Biochem., 30, 230-248 (1969).

47. W. G. Flamm, H. E. Bond, and H. E. Burr, Biochim. Biophys. Acta, 129, 310-317 (1966).

48. J. T. Lett, E. S. Klucis, and C. Sun, Biophys. J., 10, 277-292 (1970).

49. C. A. Price, E. N. Breden, and A. C. Vasconcelos, Anal. Biochem., 54, 239-246 (1973).

49a. C. A. Price and R. Casciato, Anal. Biochem., In Press.

50. H. W. Hsu and N. G. Anderson, Biophys. J., 9, 173-188 (1969).

51. P. Sheeler and J. R. Wells, Anal. Biochem., 32, 38-47 (1969).

52. T. E. Perardi and N. G. Anderson, Anal. Biochem., 34, 112-122 (1970).

53. G. B. Cline and R. B. Ryel, Methods in Enzymol., 22, 168-204 (1971).

54. N. G. Anderson, Quart. Rev. Biophys., 1, 217-263 (1968).

55. C. A. Price, in Manometric and Biochemical Techniques (W. W. Umbreit, R. H. Burris, and J. F. Stauffer, eds.), 5th ed., Burgess Publishing, Minneapolis, 1972, pp. 213-243.

56. D. H. Moore (ed.), in Physical Techniques in Biological Research, 2nd ed., Vol. 2, part B, Academic Press, New York, 1969, pp. 285-314.

57. E. F. Eikenberry, Anal. Biochem., In Press.

58. V. N. Schumaker, Adv. Biol. Med. Phys., 11, 245-339 (1967).

59. G. D. Birnie and S. M. Fox (eds.), Subcellular Components, 173 pp. Plenum Press, New York (1969).

60. G. D. Birnie (ed.), Subcellular Components, 2nd ed., 320 pp., Buttersworths, London (1972).

61. E. Reid (ed.), Separations with Zonal Rotors, University of Surrey Press, Guildford, U. K. (1971).

62. C. H. Chevrenka and L. H. Elrod, A Manual of Methods for Large-Scale Zonal Centrifugation, 85 pp., Beckman Instruments, Palo Alto, Calif. (1972).

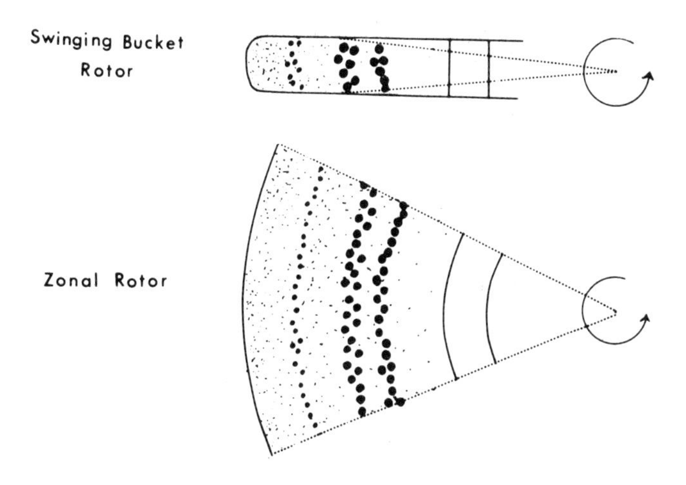

FIG. 1. Evolution of the zonal rotor. The zonal rotor can be thought of as a tube whose lateral walls have been extended through 360°. The zonal rotor thus gains an enormous increase in volume plus the elimination of "wall effect," in which particles traveling along radial lines collide with the parallel-sided walls of tubes. (Taken from Ref. 4.)

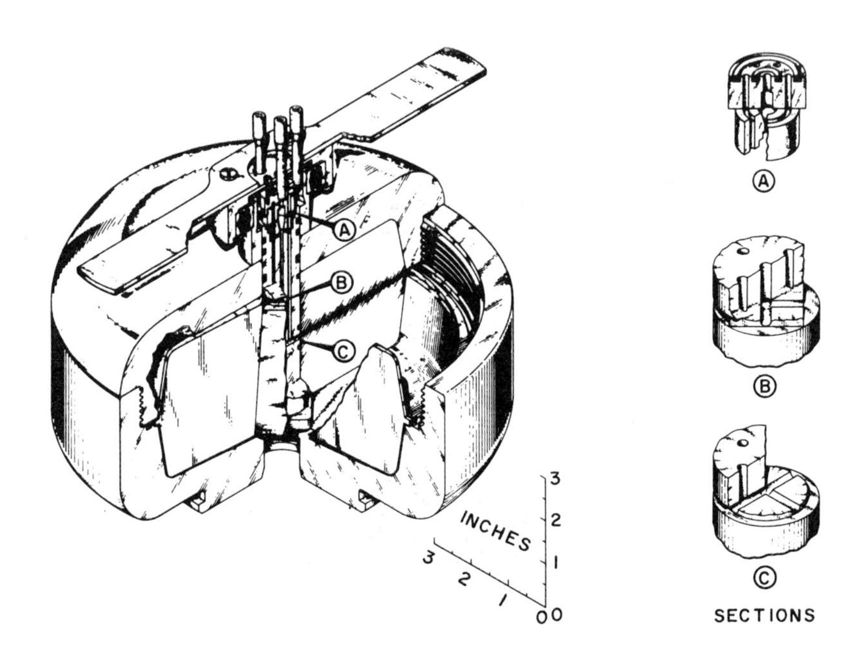

FIG. 2. Diagram of a typical zonal rotor. This cut-away view of the B-XXIX rotor shows three of the four sector-shaped compartments separated by incomplete septa. Portions of the fluid seal are shown in the inset. (Courtesy of N. G. Anderson, Oak Ridge National Laboratories.)

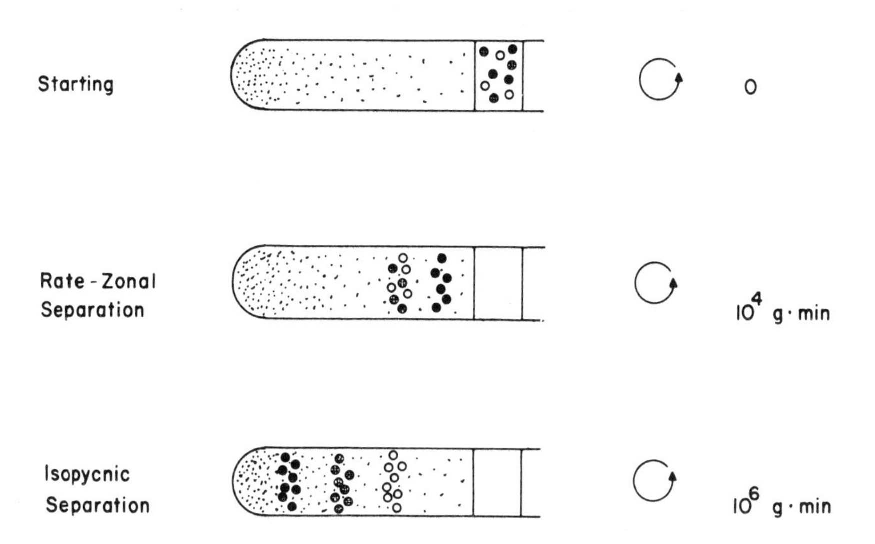

FIG. 3. Two kinds of separation in density gradient centrifugation. Particles can be separated by differences in their rates of sedimentation or, in longer or faster centrifugations, by differences in their equilibrium densities. (Taken from Ref. 55.)

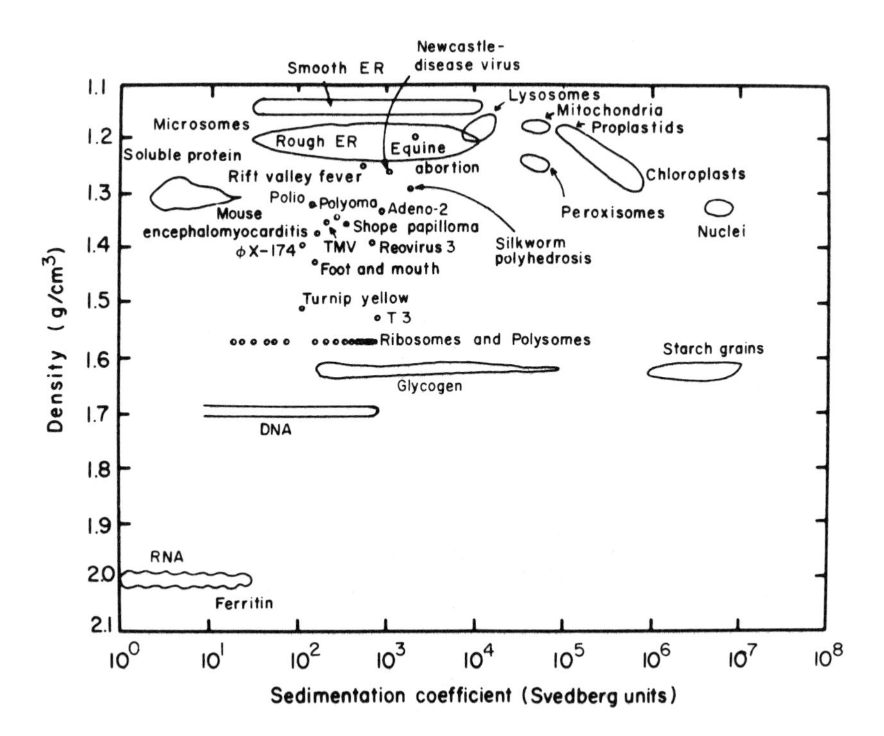

FIG. 4. S-ρ space. Anderson has pointed out that since subcellular particles tend to have unique combinations of sedimentation coefficients (S) and equilibrium densities (ρ), they can be represented by coordinates in S and ρ. (Taken from Ref. 4.)

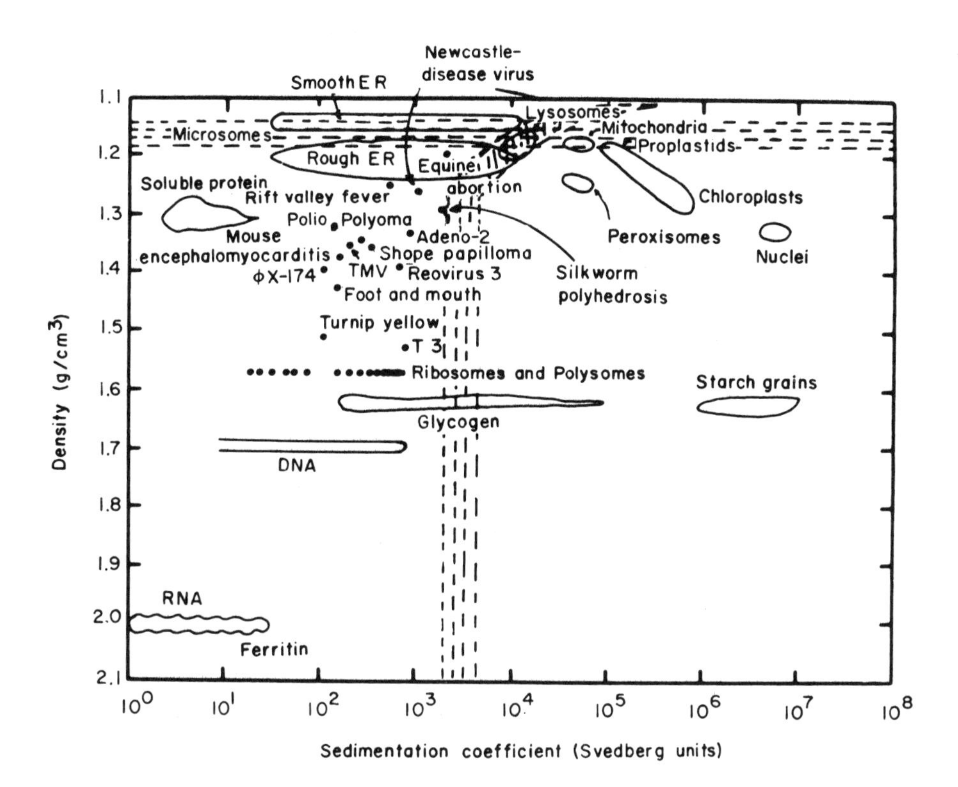

FIG. 5. Strategy of S-ρ separations. Fractions obtained from a gradient following a rate separation can be thought of as a vertical slice in S-ρ space. A second, isopycnic separation can cut these slices into horizontal cubes. (Taken from Ref. 4.)

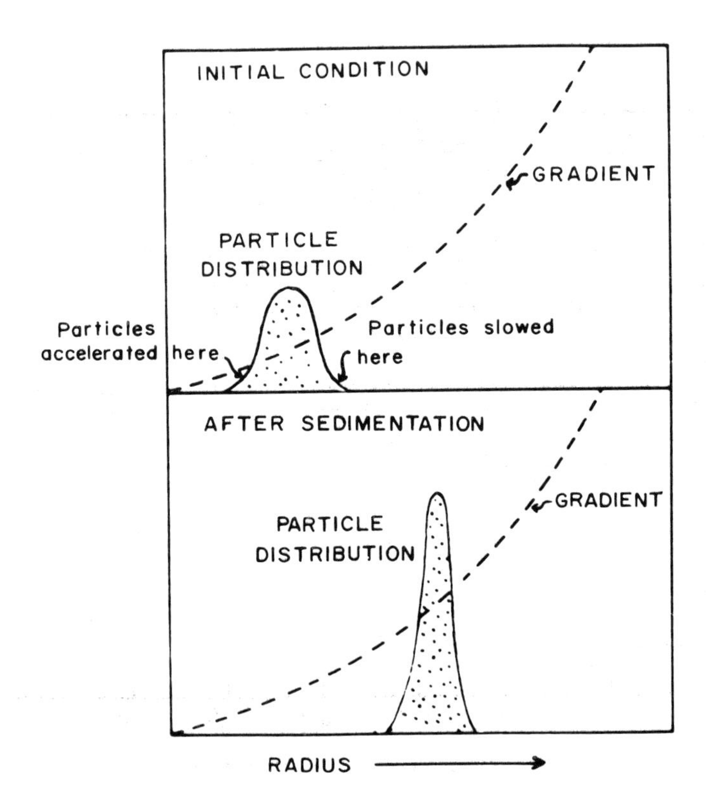

FIG. 6. Gradient-induced zone narrowing. The particles in the trailing edge of a zone in a sufficiently steep gradient will be accelerated with respect to those in the leading edge; so that the zone width will tend to decrease. (Taken from Ref. 4.)

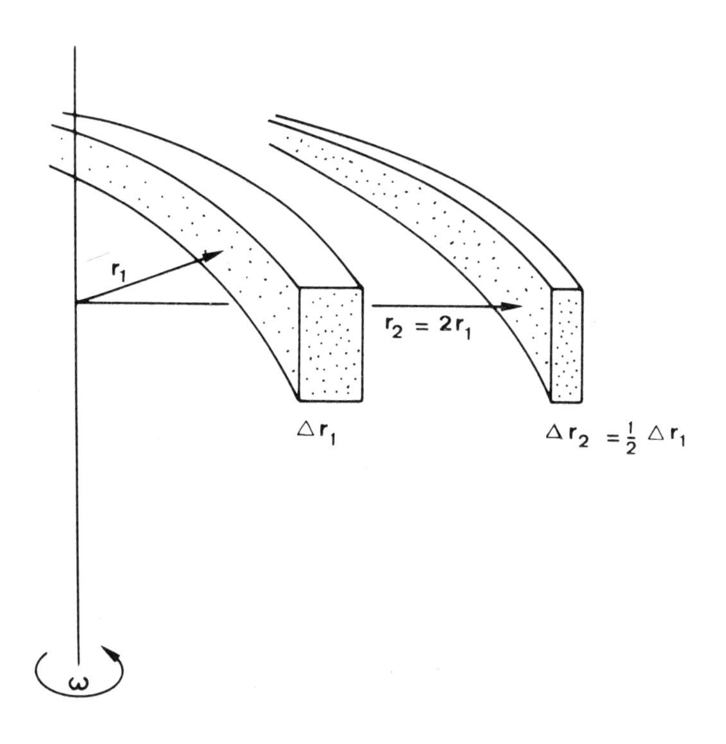

FIG. 7. Equivolumetric gradients (Eq. 4). A gradient slope can be constructed so that radial dilution is exactly counter-balanced by gradient-induced zone narrowing. (Taken from Ref. 4.)

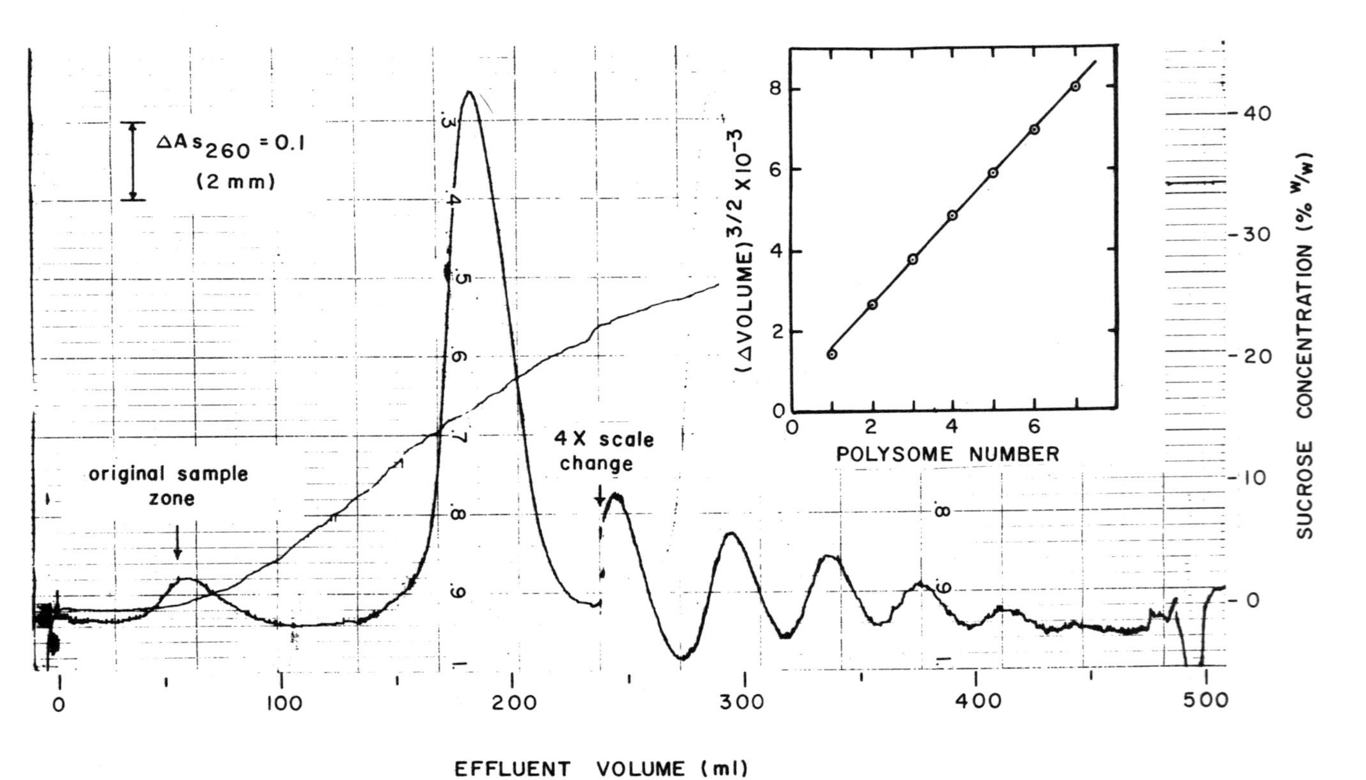

FIG. 8. Polysomes in an equivolumetric gradient. In an equivolumetric gradient, the volumetric zone widths of ribosomes of different size are independent of the distance migrated; and, as shown in the inset, the volumetric distance migrated is proportional to the sedimentation coefficient. (Taken from Ref. 8.)

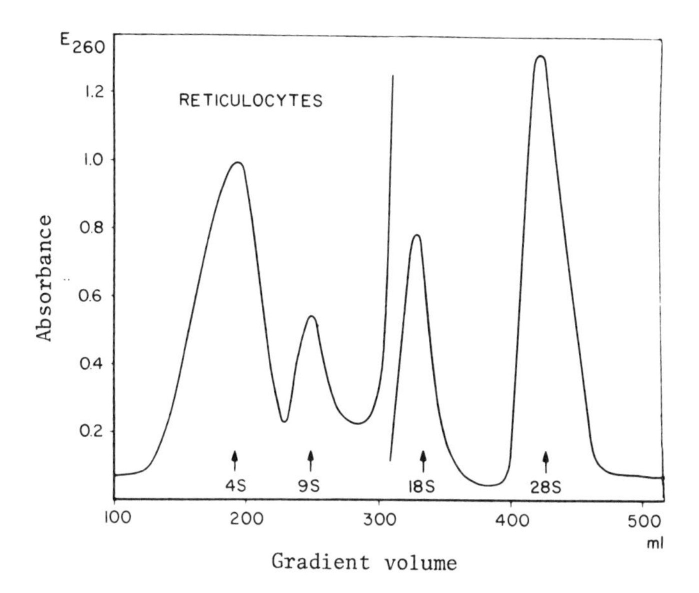

FIG. 9. Separation of RNA in an equivolumetric gradient. Reticulocyte RNA can be resolved with excellent resolution in equivolumetric gradients. (Taken from Ref. 10.)

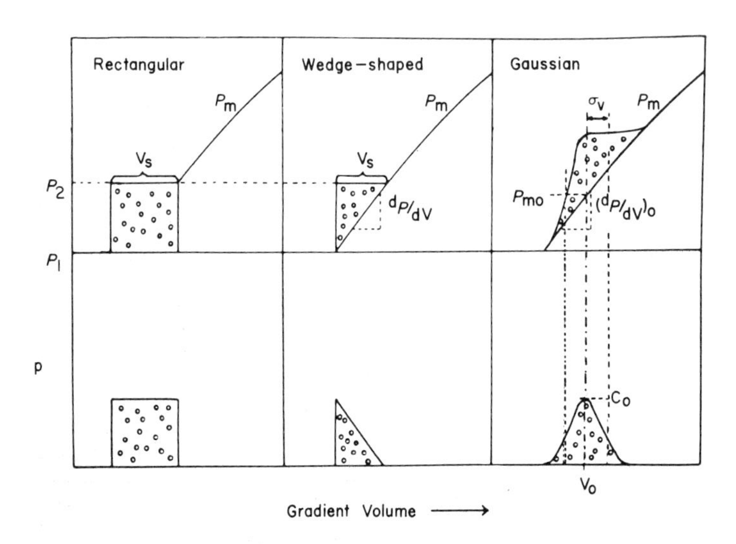

FIG. 10. Capacity of different zone shapes. The instantaneous static capacities of square, wedge-shaped and Gaussian zones are determined by the simple criterion of avoiding density inversions. The time-dependent and dynamic capacities may be substantially lower (see text). (Taken from Ref. 4.)

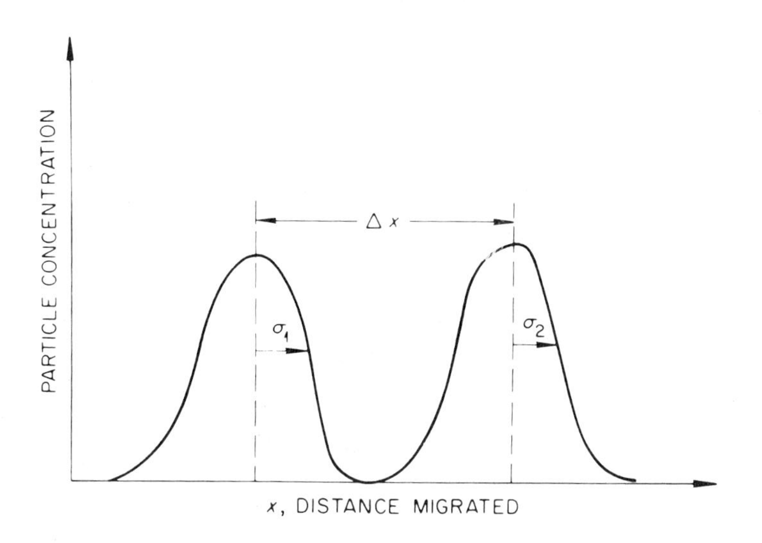

FIG. 11. Resolution in zonal rotors. Resolution can be defined as the volumetric distance between two particle zones divided by the sum of their dispersion. (Taken from Ref. 4.)

$$\text{Resolution} = \frac{\Delta X}{\sigma_1 + \sigma_2}$$

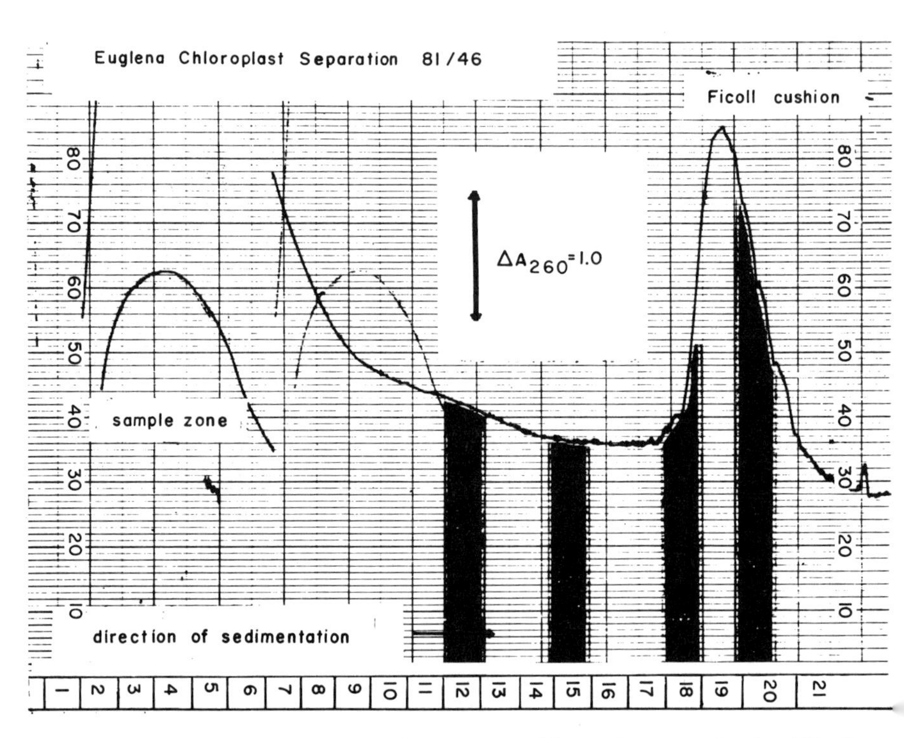

FIG. 12. Separation of Euglena chloroplasts. A clarified cell brei was subjected to rate zonal sedimentation in isomotic Ficoll in a B-30 rotor. Intact chloroplasts were recovered in the region of the gradient shown by the hatched bars. (Taken from Ref. 31.)

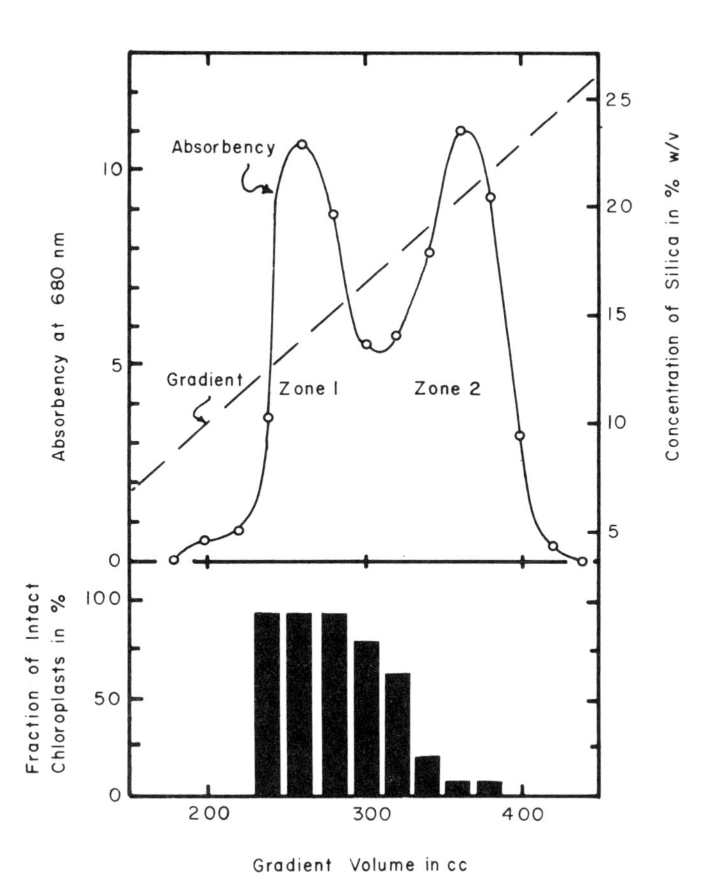

FIG. 13. Continuous-flow harvesting of spinach chloroplasts in a silica gradient. Spinach brei was passed over a silica (Ludox) gradient in a CF-6 rotor, the gradient recovered, and tested for chlorophyll (upper curve). The fractions were also examined microscopically for the intactness of the recovered chloroplasts (lower bar graph). (Taken from Ref. 49.)

FIG. 14. A simple exponential gradient. This generator, suitable for zonal rotors, can be fabricated inexpensively and is also available commercially. (Taken from Ref. 41.)

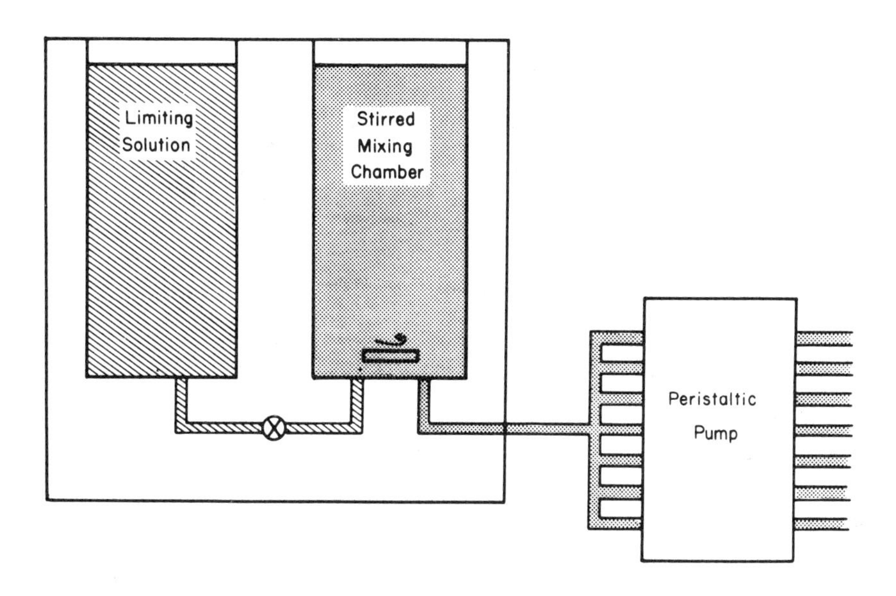

FIG. 15. Two-cylinder generator. The classic two-cylinder generator employs flow from a non-stirred into a stirred chamber, with flow out of the stirred chamber at twice the rate of flow in. For zonal rotors the volumes and viscosity of gradient material require enclosed chambers, especially vigorous mixing and usually forced flow. The diagram here shows a gradient splitter as would be employed in swinging buckets. (Taken from Ref. 4.)

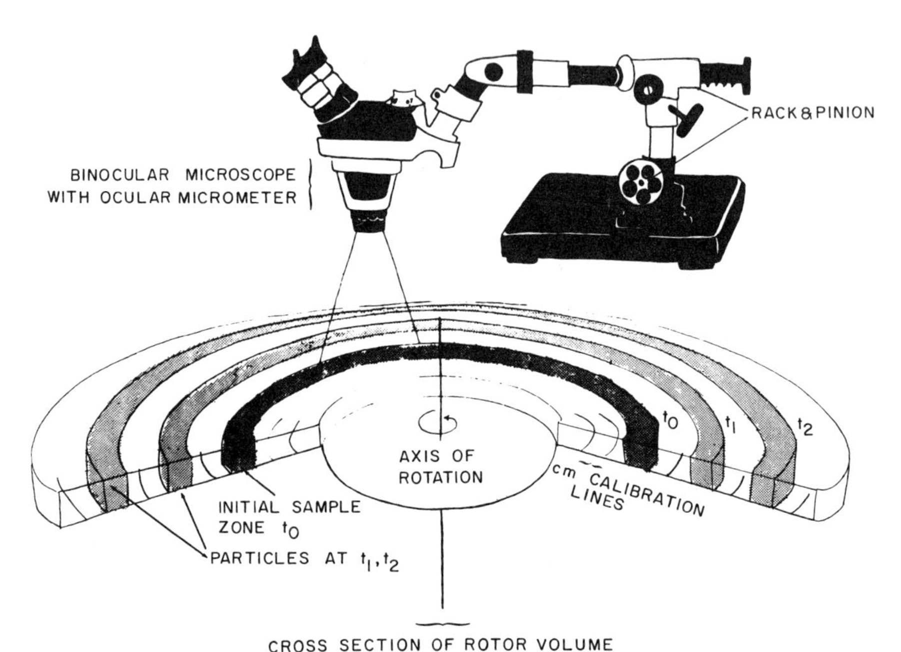

FIG. 16. Visual measurement of sedimentation rates in analytical zonal centrifugation. One may mount optical or photometric devices over transparent zonal rotors, such as the A-XII in order to measure the sedimentation of particle zones directly and continuously. (Taken from Ref. 42.)

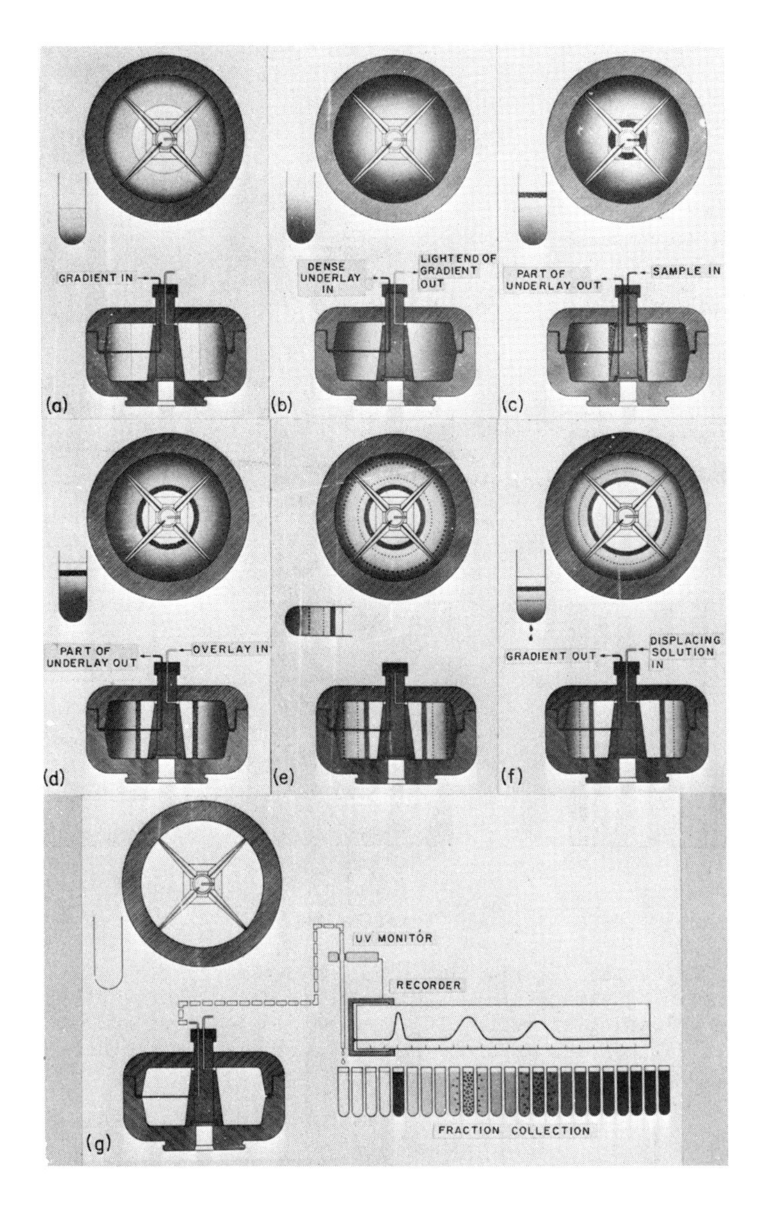

FIG. 17. Operation of zonal rotors. The steps in the operation of zonal rotors are mostly analogous to those in swinging buckets. Shown here are (a) introduction of gradient into rotor, (b) rotor with gradient in place, (c) sample in, (d) introduction of overlay, (e) particle separation, (f) recovery of gradient and particle bands, and (g) recovered gradient containing the isolated particle zones. (Courtesy of N. G. Anderson, Oak Ridge National Laboratories.)

FIG. 18. Reorienting gradient centrifugation. There are certain advantages in dispensing with fluid seals by loading and unloading at rest. The gradient is led into and out of the vertical configuration by careful acceleration and deceleration during reorientation. (Taken from Ref. 45.)

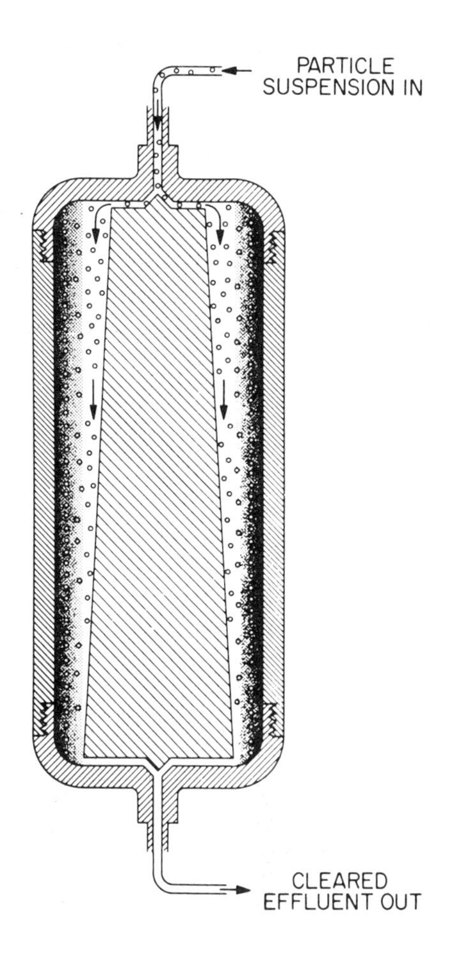

FIG. 19. K-type rotor. Continuous harvesting with isopycnic banding is practicable for virus particles with the K-type rotor. The particle suspension flows across a thin annular gradient and the spent supernatant flows out of the rotor. (Courtesy of N. G. Anderson, Oak Ridge National Laboratories.)

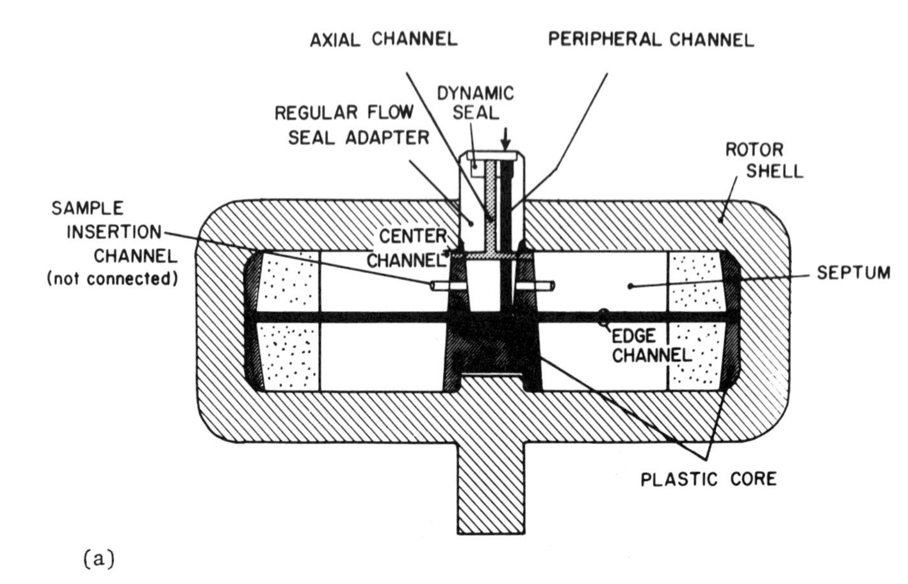

(a)

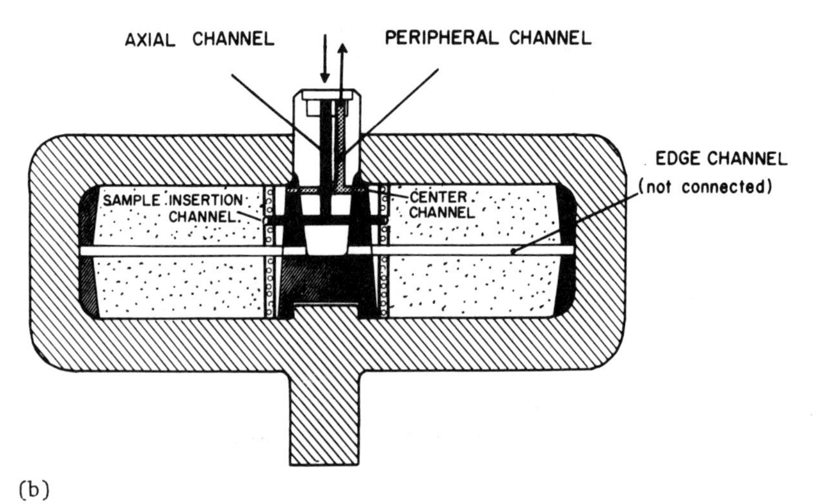

(b)

FIG. 20. MACS system for direct sample insertion. In the MACS system the seal adapter connects the seal channels to any two among a multiplicity of internal channels. In (a) the edge and center lines are connected for normal loading of the gradient; a third channel remains unconnected. In (b) a different seal adapter permits the sample to be injected through the third channel directly to a pre-determined location. Air or overlay is displaced through the center channel of the core. (Taken from Ref. 49a.)

Chapter 7

METHODS IN ELECTRON MICROSCOPY

Sydney S. Breese, Jr. and Theodore P. Zacharia

Plum Island Animal Disease Laboratory
Agricultural Research Service
United States Department of Agriculture
Greenport, New York
and
National Academy of Sciences
Washington, D. C.

I. INTRODUCTION

The advent of the electron microscope added a whole new dimension to the study of biological materials, either particulate or in tissues. The cellular level examined with light microscopy can now be extended down to the molecular level. Where many cells are examined with the light microscope, the single cell and its components are examined with the electron microscope. The techniques that are most useful in the preparation of specimens for electron microscopy and some of the detailed directions for their use will be discussed in this chapter. Furthermore, we shall attempt to show how these techniques extend the observations and help to interpret the results.

Basically, biological samples are examined in the electron microscope either as suspensions of particulates or as aggregates of cells embedded in a medium that allows thin sections to be cut and examined. Other more specialized techniques will be mentioned, but they will not be discussed in detail.

Suspensions of cells, viruses, bacteria, or cell components such as mitochondria, ribosomes, microsomes, etc., are the simplest samples to prepare for electron microscopy. They may be deposited on prepared specimen grids in various ways. The most important feature is to reduce the artifacts and contamination in the sample by use of suitable volatile suspending media.

The preparation of biological materials for thin sectioning is a more tedious, meticulous application of the arts of electron microscopy. The fixation, dehydration, embedment, sectioning, and staining of specimens is a task that requires patience and practice. Every investigator using the electron microscope should be able to perform all the necessary steps, but he will often find that the preparation of samples for research is usually done by skilled technicians. The rough details of the procedure will be given here, but study of a detailed treatise such as that of Pease [1] is necessary, along with a good deal of practice, to become sufficiently adept.

Studying electron-microscope techniques in any text is not a substitute for actually performing the procedures in a laboratory with one's own hands. The most satisfactory training comes from an apprenticeship arrangement in which the neophyte spends several months or more working at all the steps in electron microscopy in the laboratory of an experienced investigator. The next-best method is to take one of the several courses offered across the country, in which intensive work for a few weeks must be followed by practice in the laboratory. It may take six months or more to achieve results that are both meaningful and scientifically satisfying.

II. PARTICLE SUSPENSIONS

Biological specimens that consist of suspensions of particles of any kind may be mounted on the electron-microscope specimen screen by direct deposition of a very small amount with a small pipette or dropper. The suspension is allowed to remain for a few moments and is then dried quickly by touching the edge of the screen with a piece of filter paper. Alternatively, the specimen droplet may be delivered from the hole in a small bacteriological loop of platinum wire. When the suspension is such that it does not have sufficient protein concentration to spread as a thin film on the surface of the specimen screen, it is sometimes desirable to precoat the Formvar carbon surface with a film of some low-molecular-weight material such as 0.1% bovine plasma albumin in distilled, demineralized water.

Small droplets of the biological specimen may be formed also on a clean Teflon surface with or without other solutions mixed in (Fig. 1). This technique originally was developed for the visualization of DNA strands released from virus suspensions or from small quantities of purified nucleic acid [2]. Small droplets containing about 10 mg of nucleic acid are delivered by pipette on the surface of a clean Teflon baking sheet (the sizes vary in local hardware stores). The Teflon surface is hydrophobic and very spherical droplets are formed. When the original solution has contained nucleic acid or viruses that may disrupt and release their nucleic acid, each droplet forms a microdiffusion system. The surface of the droplet on the Teflon is coated with a thin film of cytochrome by a touch with a needle previously wet with a 1:1 mixture of cytochrome-powder and distilled water. The protein film is the carrier for the nucleic-acid strands that diffuse toward the surface of the droplet. Carbon-stabilized, Formvar-coated microscope screens are floated on the surface of the droplet where they pick up the cytochrome-film and the attached nucleic acid. The screens are dried by touching them to a surface of absolute ethanol and then shadow-cast with a thin film of platinum-palladium from two directions normal to each other and at a low angle or with a rotary shadowing device. Each droplet may be used for a single observation, but there is room for many droplets on the Teflon sheet and many variations can be made in the conditions, concentrations, and time of diffusion. Of course, this method is not limited to the study of nucleic acid. It may be used with any solutions that might be examined in the electron microscope; many variations have been used.

The suspension of particles may also be mounted on the specimen screen in such a way that they may be counted quantitatively. This may be done by spraying and counting methods, by use of mixtures of the unknown suspension and known dilutions of a standard particle such as Dow latex beads [3]. Spray droplets are formed by the impingement of a larger drop on the glass button in the nasal spray device shown in Fig. 2.

The proportions of known concentration of Dow polystyrene latex and virus may vary in individual experiments. The Dow latex particles obtained from Dow Chemical Co., Midland, Michigan, come in several sizes, the most useful being LS-040-A (88 nm diameter), LS-055-A (188 nm diameter), and LS-057-A (264 nm diameter), which in our samples contained 31 x 10^{13}, 3.0 x 10^{13}, and 1.0 x 10^{13} particles per ml, respectively. When used in a typical spray-droplet-counting experiment with a virus suspension containing 4.8 mg/ml, the following mixture was put in the spray gun: 0.2 ml of a 1/1000 dilution in distilled water of the virus suspension, 25 μl of Dow latex (88 nm) diluted 1/400 in distilled water, and 25 μl of 1% bovine plasma albumin in distilled water. When the virus concentration gets below 10^7 particles per ml, the chances for accurate virus counts become very slight, even when particles are counted in many droplets. For low concentrations of particles, Sharp [4,5] has developed a method using centrifugation and collection on agar blocks. The agar surface is then psuedoreplicated with parlodian solution and examined in the electron microscope.

Two methods are used to delineate the structure of particle suspension, shadowing and negative staining. Shadow casting (Fig. 3) is done in a specialized vacuum chamber in which the support for the specimen screens may be varied in angle and position with respect to the source of volatized metal or carbon. The metals used may be gold, uranium, chromium, and various alloys of palladium and platinum. All of these provide a high-contrast shell on the side toward the metal source and a clear uncoated shadow on the opposite side, as may be seen in the diagrams of Fig. 3. Negative staining or positive staining may be done by addition of certain highly electron-scattering solutions to the suspension of particles or by addition of the solutions as a second step after the suspension has dried on the support film [6]. In each case, a loop droplet or floatation of the screen, film-side down, on a droplet of suspension is used to make the specimen. The most commonly used solutions are phosphotungstic acid, phosphomolybdic acid, silicotungstic acid, uranyl acetate, uranyl nitrate, and cadmium iodide. The concentrations may vary from 1-2% for the three acids, up to saturated solutions for the uranyl compounds. The difference in density between the biological material and the chemical solution is used to great advantage to enhance the contrast between specimen and background, and to take advantage of the resolution of the electron microscope.

III. ULTRATHIN SECTIONS

The technique for biological specimens that has proved the most useful since its development in the early 1950's is ultrathin sectioning. The specimen must be fixed, dehydrated, embedded, sectioned, and stained. Each one of these steps had to be developed and adapted to the peculiar requirements of the electron microscope. The specimen size, first of all, must be small. Bits of tissue or pellets of particles must be less than 1 mm on a side during the processing, and are even smaller when trimmed for the actual sectioning.

The steps for epon embedding will be given in some detail, but other embedding materials may be advantageous in specific cases [7]. The necessary materials are as follows.

1. 0.1 M phosphate buffer with 1% sucrose (pH 7.2-7.5), prepared with:

   $NaH_2PO_4 \cdot H_2O$, mol wt 138.01, 13.80 g/l of H_2O;

   Na_2HPO_4, mol wt 141.96, 14.20 g/l of H_2O;

   0.1 M $NaH_2PO_4 \cdot H_2O$, 20 ml;

   0.1 M Na_2HPO_4, 80 ml;

   sucrose, 1 gH.

2. 1% glutaraldehyde;

   0.1 M phosphate buffer with 1% sucrose, 24 ml;

   glutaraldehyde, 25% aqueous, 1 ml.

3. 1% OsO_4 (osmium tetroxide, osmic acid);

   0.1 M phosphate buffer with 1% sucrose, 25 ml;

   OsO_4, 1/4 g.

Be very careful to observe all precautions with OsO_4 and prevent exposure of eyes, nose, or face to vapors. Open ampoules in a hood.

4. Epon resin mixture: DDSA--dodecenyl succinic anhydride; MNA--nadic methyl anhydride; DMP-30--2,4,6-tri(dimethylaminoethyl)phenol.

   Mix A = 62 ml Epon 812 + 100 ml DDSA

   Mix B = 100 ml Epon 812 + 89 ml MNA

   A/B Epon = 80 ml Mix A + 20 ml Mix B + 2 ml DMP-30

All the above must be thoroughly mixed with a Teflon stirrer in glass beakers. The mixtures must be complete and uniform since each of the ingredients is very viscous.

5. Ethyl alcohol concentrations in distilled H_2O: 30%, 50%, 70%, 90%, Absolute.
6. Propylene oxide--store at 4°C until used.
7. Capsules, either gelatin size 00 or plastic (BEEM).

The method given here has proven very acceptable for pellets of tissue-culture cells and also for small blocks (less than 1 mm on a side) of tissues from animals. The primary fixation is for 30-60 min in cold (4°C), fresh 1.0% glutaraldehyde. When tissues are involved, the pieces are minced in the glutaraldehyde and then collected and put into fresh glutaraldehyde solution for another 30 min. Tissue-culture cells may be washed in the bottle with phosphate buffer and then scraped into a few milliliters of cold glutaraldehyde. They are then poured into a conical centrifuge tube and sedimented for 3-5 min at about 700-800 x g. The pelleted cells are held in the cold (4°C) for 30-60 min and then washed 3 times with phosphate buffer.

After glutaraldehyde treatment and washing, the samples are suspended in 1% osmium tetroxide in buffer and held at 4°C for 30-60 min. The samples are then washed again with phosphate buffer at least 3 times, with good resuspension each time to facilitate the removal of excess osmium.

The suspensions are then dehydrated by passage through the cold-graded alcohols. The alcohol is decanted from the tubes and the next-higher concentration is added until 90%. The tubes are then allowed to warm to room temperature. Three changes of 100% ethyl alcohol are followed by two changes of propylene oxide.

The last wash with propylene oxide is replaced by a 1:1 mixture of propylene oxide and A/B Epon with DMP-30 for 30 min. This is followed by transfer into the embedding mixture of A/B Epon and DMP-30 for another 30 min. The final transfer takes place into capsules filled with the Epon-embedding mixture. At this stage, just one or two tiny pieces of sample are placed in each capsule, where they slowly sediment to the tip of the capsule. The capsules are topped off with Epon and closed before polymerization.

The polymerization takes place during about 20 hr at 60°C. If, when that period is over, the Epon appears soft, a higher temperature may be used for a few hours to further harden the mixture.

The solutions and mixtures for this embedding procedure may all be made in advance and stored at 4°C. The mix A and mix B

Epons must be brought to room temperature before being combined to make the embedding mixture. The final mixture with DMP-30 is prepared while the preceding steps are under way. Make only enough for immediate use, because the DMP-30 is a hardening and polymerizing agent and will act even at 4°C. Small labels indicating the contents should be made and placed in each capsule when the Epon is added. This assures identification of the embedment at any later time when it may be cut. The Epon preparations may be varied in hardness and toughness so that they will both infiltrate the particular biological speciment to be examined and allow good cutting qualities besides.

Thin sections are cut with specially designed microtomes by use of glass or diamond knives. The microtomes are made by several companies and each one has features that are preferred by individual investigators. Diamond knives are expensive, but in the long run provide more consistent results than glass knives. Every microscopist must start with glass knives first before risking the edge on a diamond knife. The economy of glass is best when students are learning the techniques and have many specimens for practice.

IV. SCANNING ELECTRON MICROSCOPY

The technique of scanning electron microscopy is rapidly coming into prominence in the study of both biological and nonbiological specimens. Albert Crewe [8] has pointed out that in principle the transmission electron microscope and the scanning microscope are very similar. Figure 4 makes the comparison, with the scanning microscope diagram drawn with the source of electrons at the bottom, so that the positioning of lens, aperture, and specimen are identical in the two systems. On the left, the source of electrons illuminates the specimen and the image is formed by the magnetic-lens system in a point-to-point fashion. The right-hand diagram shows how the magnetic lens focuses the beam of electrons on the specimen and is continuously moved back and forth by the scanning coils. The result is the release of energetic particles from the specimen, that in turn are collected at a detector and translated into an image.

The basic structure of the advanced microscope developed by Crewe [8] is shown in Fig. 5. By use of a field-emission tip for a high-intensity stream of electrons focused by the aperture and lens system, the specimen is scanned with a very small beam with high electron density. This system results in higher resolution than with the present scanning microscopes. The image-detection system collects transmitted electrons which pass through an energy analyzer before being registered in the detector and displayed on a cathode-ray tube. A simultaneous display on a second cathode-ray tube is photographed with a Polaroid film for a permanent record.

Commercial scanning electron microscopes now operate at resolutions of 10-25 nm (100-250 Å) while the Crewe instrument has a resolution of about 0.5 nm (5 Å), comparable to the best transmission microscopes. The receiving modes of the scanning instruments may be activated by secondary, back-scattered or transmitted electrons as well as x-ray or luminous fluorescence. This means that the scanning microscope will be able to supplement the information of the regular transmission microscope and with the Crewe instrument at the same level of resolution. Biological specimens generally must be provided with an electron-conducting coating that is put on by rotation of the specimen in a vacuum evaporator that deposits gold or palladium alloy. The scanning microscope accepts much larger specimens than does the transmission instrument. The fact that the specimen may be rotated and tilted as well as the fact that the angle of the beam of electrons may be varied provides very exciting three-dimensional views of the surfaces scanned.

V. SPECIALIZED TECHNIQUES

With the advent of some specialized techniques that can be extended from the light-microscope level to the electron-microscope specimen, very precise and delicate cytological and biochemical studies can be made. The techniques that have had the greatest application have been autoradiography, immuno-electron microscopy, ultrastructural cytochemistry, enzyme cytochemistry, and high-resolution studies. The application of these techniques is very exacting in most cases and requires that the investigator have a very good command of the basic electron microscopical skills. The difficulties generally come from the need to handle the specimens many times and to have many controls and preliminary specimens to aid in the final interpretation of results.

The development of autoradiographic techniques for the electron microscope has evolved over the years, starting with Caro [9] and continuing toward more quantitative accuracy by Salpeter [10]. The basic problem was to assure that the specimen of tissue, the overlying emulsion, and the supporting plastic film were thin enough to be penetrated by the electron beam. The specimens for autoradiography are tissues or tissue-culture cells that have been exposed to chemical compounds containing ^{3}H. The electrons emitted from tritiated compounds have the most desirable characteristics for autoradiography. Figure 6 shows the cross section of a typical specimen. The tracks of the β^{-} particles are shown to be going in several directions. When there is an interaction between the emitted electron and silver bromide in the emulsion, a silver grain is deposited that will later be visible in the electron microscope. The cells that have absorbed the radioactive compound are fixed, postfixed, and embedded in the usual fashion (Fig. 7). The Epon sections are cut in ribbons, 4 or 5 of which are then gently transferred to a small drop of water on a glass microscope slide previously coated with a thin film of collodion. The water droplet is removed, and the short ribbons dry down on the collodion. Staining with uranyl acetate and lead can be done at this stage, followed by stabilization of the sections by evaporation of about 5 nm (50 Å) of carbon on them. The slide is then coated with a liquid emulsion. Several emulsions are used, with Ilford L4 and Kodak NTE being the most easily applied by the beginner. When the emulsion is dry, the slides are stored in a cool dark box until the reactions have had time to produce visible clusters of silver grains. The glass slides are subsequently carried through the photographic development and fixation process. The last step is to float the collodion film from the glass slide and then to pick up the individual ribbons on microscope support screens. As might be expected, the processing, involving the preparation, application, and development of the liquid emulsion, must be done under suitable darkroom conditions.

In order to have successful experiments, it is necessary to have many preliminary specimens for determination of the effective amounts of radioactivity and to test the thickness of collodion, carbon, and emulsion, all of which go into each specimen. The detailed techniques are given by Salpeter [10]. More sophisticated specimens may be interpreted to give quantitatively accurate results as to the location of radioactive centers [11].

The study of immunological phenomena can be enhanced by the use of labeled antibodies to react in cellular areas that cannot be seen in the light microscope. The most effective method has been to attach purified ferritin to the antibody molecule at one site, while leaving free other sites that may react with specific antigens in the biological samples. In this technique, considerable effort is required to purify horse-spleen ferritin by successive crystallizations with 5% cadmium sulfate. The purified ferritin is then cross-linked to antibody globulin with meta-xylene di-isocyanate. In the experimental system, a cell containing the antigen or a tissue-culture system infected with a virus, for example, is treated with ferritin conjugate and then washed thoroughly to remove the unused material. The cells are embedded in the standard way and the thin sections are examined in the electron microscope. The successful experiment shows the antigen surrounded by the very-electron-dense small particles of ferritin, in which the iron cores are scattering the electrons. The techniques in immunoelectron microscopy have been applied in cancer, virus, organ transplant, and immunochemical research. The use of fluorescent antibody to determine sites of antigen formation is well known in light microscopy. Ferritin tagging allows the search to be extended to the individual cells. Other immunoelectron microscopy techniques have involved uranium and mercury as heavy-metal labels on antibodies, but they have not been as easily adaptable as ferritin. A recent detailed explanation of the entire ferritin conjugation technique has been published [12].

Ultrastructural cytochemistry involving the selective enzyme digestion of biological structures mounted in thin sections has been useful in work with viruses and certain cellular elements. With this technique, developed by Bernhard and Tournier [13], the biological specimens are embedded in a water-soluble medium such as glycol methacrylate. The specimen is first lightly fixed with formalin or glutaraldehyde and then dehydrated with embedding plastic. The tissues are not treated with osmium and the plastic retains enough moisture to let the enzyme solution work on the thin section. The sections are floated on the appropriate enzyme for various lengths of time for determination of the optimum effect. Figure 8 diagrammatically shows the results of the use of pepsin to remove the protein coat from either reovirus or adenovirus, as well as the selective action of DNase and RNase to remove the nucleic-acid cores. The DNase removes the adenovirus core while not affecting the reovirus, while RNase removes the reovirus center but leaves adenovirus untouched. The effects of the enzyme digestion are more difficult to evaluate when the virus is smaller in diameter than the thickness of the section, as shown in Fig. 9. The results on the section on the left are much more apparent than those on the right, where the virus does not extend through the plastic and the enzymes do not have as much material to work on.

The location of sites of enzyme activity in cells is another special technique of electron microscopy. The technical details are more complicated because of the smaller samples and multiple handlings required. The tissues must retain the enzyme activity throughout the fixation, treatment, and embedment processing. For enzyme studies, small blocks of tissue are frozen or fixed in glutaraldehyde or other aldehyde fixatives. They are then treated with a reaction mixture that is specific for the enzyme and that, through a secondary reaction, may be transformed into an insoluble compound of high electron density. The materials must be such that they do not migrate during the processing for thin sections. Other methods include the use of modified azo dyes that are slightly low in density electrons, and applications of certain organic compounds that will subsequently bind to OsO_4 and provide high electron density. A good review has been written by Holt and Hicks [14].

Other specialized techniques have been developed with the idea of providing higher resolution. One of the most effective, of course, is to be meticulous in the alignment and adjustment of the microscope. It is particularly important to devote much attention to cleaning the instrument and the specimen holder and then checking the adjustment of the objective lens and the apertures. Higher-voltage microscopes are a step toward the future and, before long, there will be regional centers where microscopes operating at up to 1 million volts will be available for special studies. As higher resolution is approached, the interaction of specimen and support becomes more important. Furthermore, with higher vacuum and operation of lenses at ultralow temperature, the problems of contamination are greatly reduced and resolution is improved. The latest and most successful experimental microscope [15] has combined high vacuum, low temperatures, and lenses operated by superconducting circulating currents at liquid helium temperatures.

Another method that has been developed in recent years is freeze etching. In this method, small samples of tissue or cells are frozen at liquid nitrogen temperatures and fractured inside a vacuum bell jar. The fractured surfaces are then coated and replicated in the same vacuum system. When the vacuum is broken, the original material is removed from the replica and viewed in the electron microscope. By extremely careful work, Steere and co-workers [16-18] have been able to show both surfaces of the fracture. From these surfaces it is possible to make new evaluations and interpretations of the cellular organelles. It takes considerable practice to recognize just where the fraction has taken place. Because the technique requires replication of a surface, it does not provide the highest resolution. However, the unique views of the interiors of cells make the micrographs very exciting.

VI. NOTES ON THE INTERPRETATION OF ELECTRON MICROGRAPHS

A very difficult task for the electron microscopist is the interpretation of micrographs for a specific experiment. To analyze the situation, we must first consider the dimensions that apply and the sampling process that all electron microscopy involves. Take, for example, a tissue culture in a 4-oz prescription bottle that may have some 4-5 millions cells as a monolayer. When these cells are then infected with perhaps 0.5 million virus particles, the virus replicates and, after a certain period of time, a large number of cells are infected. When these cells are collected, washed, fixed, embedded, and sectioned, the individual section seen in the microscope represents only a few micrometers in dimension. In the usual situation, enough virus has been formed in the cells so that examination of a few cells or several ribbons of sections will provide areas in which virus may be found. In general, it is easier, of course, to find a large virus than a small one and to find aggregates of virus rather than single ones. When the concentration of virus gets below about 1 million per milliliter in the infected culture, it becomes more and more difficult to find the virus in the electron microscope sections.

The next step would be to increase the number of samples examined. Obviously, this becomes difficult mainly in terms of time and effort. Preparing several hundred sections and looking at each of them require much more time perhaps than trying to redesign the experiment to increase the number of virus particles per cell. These considerations must be taken into account when the results of an electron microscopic study are examined.

The examination of particles or viruses that have been treated with heavy-metal stains may lead to many difficulties in interpretation. The penetration of the particles and the amount of stain in small holes and interstices are vital in forming conclusions about the structure of the particular virus. Several methods have been devised to aid in the interpretation. One of these is the rotation method of Markham [19] in which the enlarged image of the virus particle is rotated stepwise in a circle to eliminate random structural detail while, at the same time, emphasizing the basic repeating structure in the particle. This method unfortunately also has its faults, and others have been devised to help particularly with virus structure. The most sophisticated has been the work of Klug, Finch and collaborators [20] in which plastic construction set models have been used to duplicate the virus structure. By special lighting and photographic manipulation, the models and the electron micrographs are reconciled to give a very accurate interpretation of structure.

Based on tissue samples which have been well fixed and stained with uranyl acetate and lead citrate, the structures of cell components may be recognized by reference to Table 1 [21].

TABLE 1

Recognition of Cell Components

Cellular element	Appearance after good fixation
Mitochondria	Few swollen or empty looking; external and internal membranes are smooth, no gross distortion in shape; clear cristae.
Endoplasmic reticulum	Membranes intact, essentially parallel; when stacked, cisternae arrangement is uniform; polyribosomes are clear in rough endoplasmic reticulum.
Golgi membranes	Intact, vesicles of various dimensions appear in a circumscribed area.
Plasma membrane	Intact and smooth around entire cell, continuous with various membrane invaginations and evaginations.
Cytoplasmic matrix	Fine precipitate; no empty spaces are generally seen.
Nuclear envelope	Both membranes are intact (except at pores) and essentially parallel to each other.
Nuclear contents	Finely granular with denser masses both in the interior and periphery of the nucleus.

The interpretation of results using the more complicated techniques of ferritin tagging and autoradiograph is even more difficult. The problem is basically one of experimental design in which it is necessary to allow for many possibilities and to make control specimens for each step of the experiment. The addition of each new item must be accompanied by sections in which the tissue is the same but the ingredient is missing. Such care will show how the experiment has progressed and will save a good deal of time in the long run.

Thin sections are judged for thickness by the use of the following criteria:

Thickness	Interference color
10-60 nm (100-600 Å)	Gray
60-90 nm (600-900 Å)	Silver
90-150 nm (900-1500 Å)	Gold
150-190 nm (1500-1900 Å)	Purple
190-240 nm (1900-2400 Å)	Blue

In autoradiography the total thickness, including the 5-nm (50-Å) carbon layer, photosensitive emulsion, and specimen (see Figs. 6 and 7), changes the color, and means that the entire preparation should not exceed 150-190 nm (1500-1900 Å) or purple interference color. The work is also done in the darkroom where the eye has to adjust to estimating the thickness without the natural color being visible.

In ferritin-tagging experiments nonmicroscopic techniques such as immunoelectrophoresis or agar gel diffusion should be used to check on the efficacy of the ferritin conjugation. Both antigen-antibody reactions and the reaction of the ferritin against rabbit antiferritin serum can be checked by these methods [12].

Unfortunately, any remarks that are made about interpretation of micrographs cannot substitute for experience. The whole subject of interpretation of ultrastructure has been dealt with in a volume edited by Harris [22]. It is only when an investigator has examined specimens in the microscope and has examined the micrographs of his own experiments and those of others that he can become confident in interpretation. As in any field that combines art and science, it is this continuous learning process with practice that makes for a skilled electron microscopist.

REFERENCES

1. D. C. Pease, Histological Techniques for Electron Microscopy, 2nd ed., Academic Press, New York, 1964.

2. H. D. Mayor and L. E. Jordan, Science, 161, 1246 (1968).

3. S. S. Breese, Jr. and R. Trautman, Virology, 10, 57 (1960).

4. D. G. Sharp, Fourth International Congress on Electron Microscopy (W. Bargmann, D. Peters, and C. Wolpers, eds.), Springer-Verlag, Berlin, Gottingen, Heidelberg, 1960, p. 542.

5. D. G. Sharp, Lab. Invest., 14 (6), pt. 2, 831 (1965).

6. R. C. Valentine and R. W. Horne, The Interpretation of Ultrastructure (R. J. C. Harris, ed.), Academic Press, New York, 1962, p. 263.

7. J. H. Luft, J. Biophys. Biochem. Cytol., 9, 409 (1961).

8. A. V. Crewe and J. Wall, J. Mol. Biol., 48, 375 (1970).

9. L. G. Caro, Methods in Cell Physiology, Vol. I (David M. Prescott, ed.), Academic Press, New York, 1964, p. 327.

10. M. M. Salpeter, Methods in Cell Physiology, Vol. II (David M. Prescott, ed.), Academic Press, New York, 1966, p. 229.

11. M. M. Salpeter, L. Bachmann, and E. E. Salpeter, J. Cell. Biology, 41, 1 (1969).

12. S. S. Breese, Jr. and K. C. Hsu, in Methods of Virology, Vol. 5 (K. Maramorosch and H. Koprowski, eds.), Academic Press, New York, 1971.

13. W. Bernhard and P. Tournier, Cold Spring Harbor Symp. Quant. Biol., 27, 67 (1962).

14. S. J. Holt and R. M. Hicks, The Interpretation of Ultrastructure (R. J. C. Harris, ed.), Academic Press, New York, 1962, p. 163.

15. H. Fernandez-Moran, Microscopie Electronique, Vol. 2 (P. Favard, ed.), Societe Francaise de Microscopie Electronique, Paris, 1970, p. 91.

16. R. L. Steere and M. Moseley, Microscopie Electronique, Vol. 1 (P. Favard, ed.), Societe Francaise de Microscopie Electronique, Paris, 1970, p. 451.

17. R. L. Steere, Proceedings 29th Annual Meeting, Electron Microscopy Society of America (Claude J. Arcenaux, ed.), Claitor's Publishing Division, Baton Rouge, Louisiana, 1971, p. 242.

18. A. E. Demsey and R. L. Steere, Proceedings 29th Annual Meeting, Electron Microscopy Society of America (Claude J. Arcenaux, ed.), Claitor's Publishing Division, Baton Rouge, Louisiana, 1971, p. 440.

19. R. Markham, S. Frey, and G. J. Hills, Virology, 20, 88 (1963).

20. A. Klug and J. T. Finch, Journ. Mol. Biol., 11, 403 (1965); A. Klug, Ibid., 11, 425 (1965); J. T. Finch and A. Klug, Ibid., 15, 344 (1966); J. T. Finch, A. Klug, and R. Leberman, Ibid., 50, 215 (1970).

21. S. Wischnitzer, Introduction to Electron Microscopy, 2nd ed., Pergamon Press, Elmsford, New York, 1970, p. 129.

22. R. J. C. Harris (ed.), The Interpretation of Ultrastructure, Vol. I, Academic Press, New York, 1962.

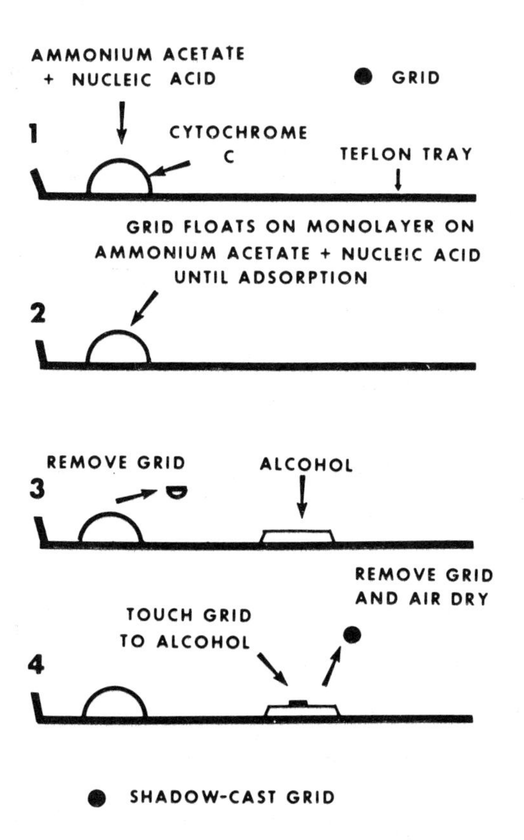

FIG. 1. The steps necessary to prepare monolayer films on small droplets of substrate on a Teflon tray. From [2]: Fig. 1 by permission.

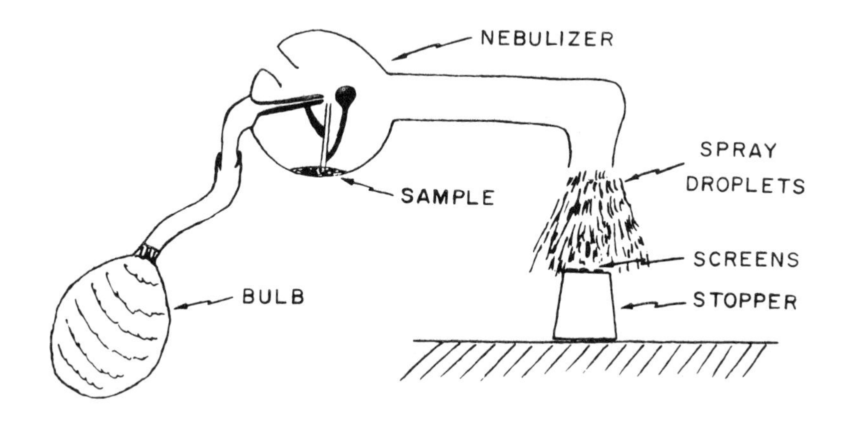

FIG. 2. Apparatus for producing low-velocity spray droplets. From [3]: Fig. 1 by permission.

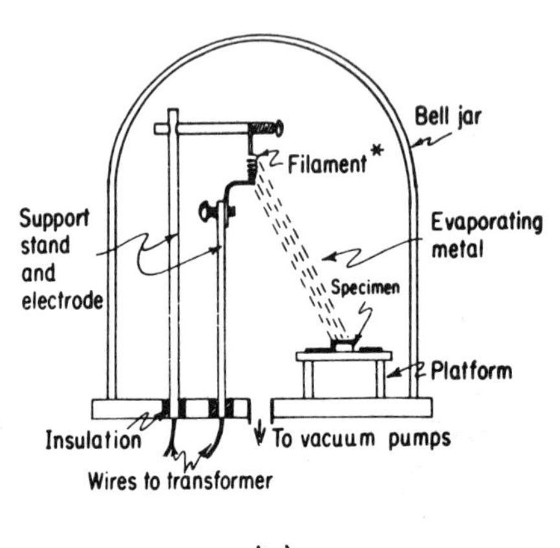

(a)

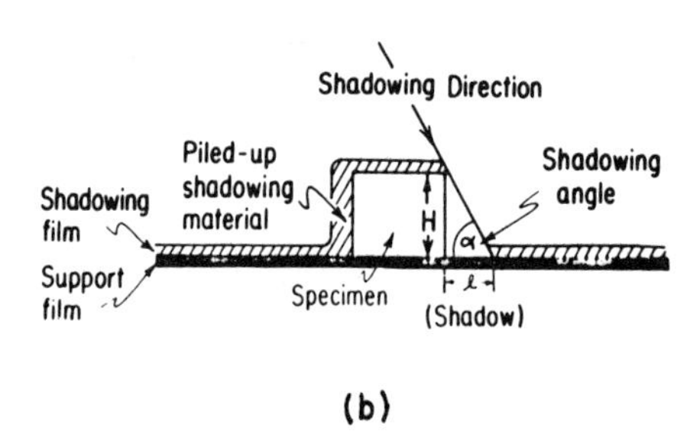

(b)

FIG. 3. (a) Diagram of standard shadow-casting apparatus with bell jar, showing heated tungsten filament containing the metal to be evaporated. (b) A higher-magnification view of metal deposited on specimen and creation of a "shadow." From [19]: p. 244 by permission.

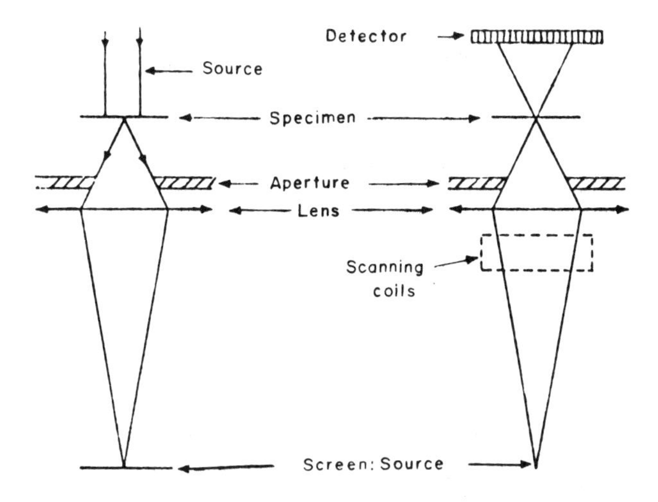

FIG. 4. Comparative diagrams for the transmission (left) and scanning (right) electron microscopes to show the similarity in principle of design. From [8]: p. 377 by permission.

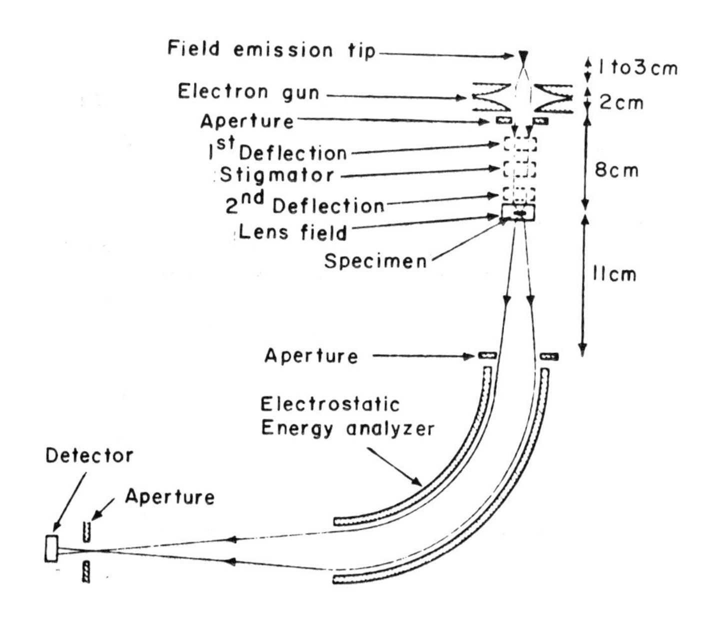

FIG. 5. The schematic diagram of the high-resolution scanning electron microscope. From [8]: p. 383 by permission.

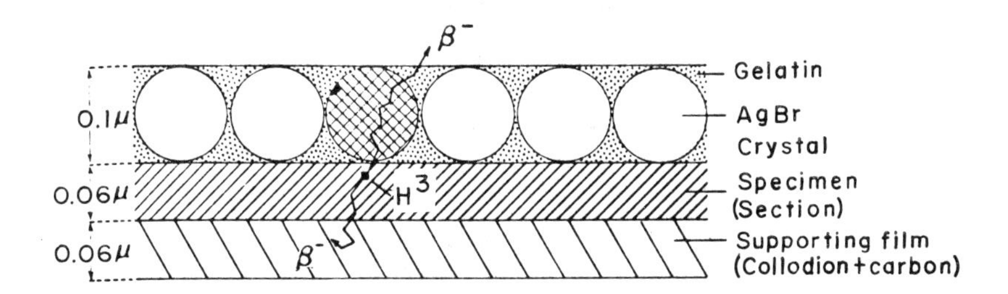

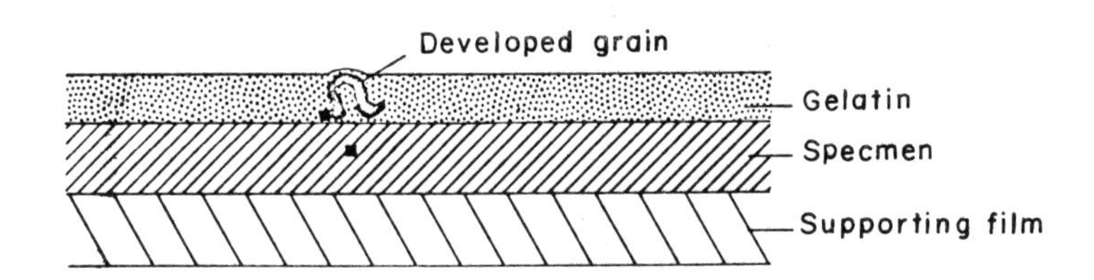

FIG. 6. The cross section of a typical autoradiography sample with the emission of electrons from a source. The lower diagram shows the resulting developed grain seen superimposed with the source in the electron microscope. From [9]: Fig. 1 by permission.

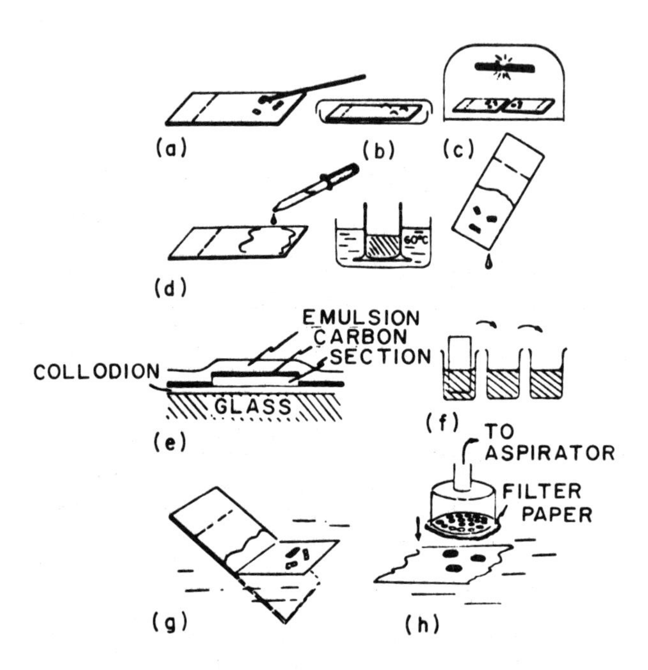

FIG. 7. The steps in preparation of an autoradiographic specimen. (a) mounting sections, (b) staining, (c) carbon stabilization, (d) coating with liquid emulsion, (e) total specimen before storage for reaction time, (f) developing and fixing, with washing in between, (g) floating-off sections, and (h) collecting on grids by suction. From [10]: Fig. 8 by permission.

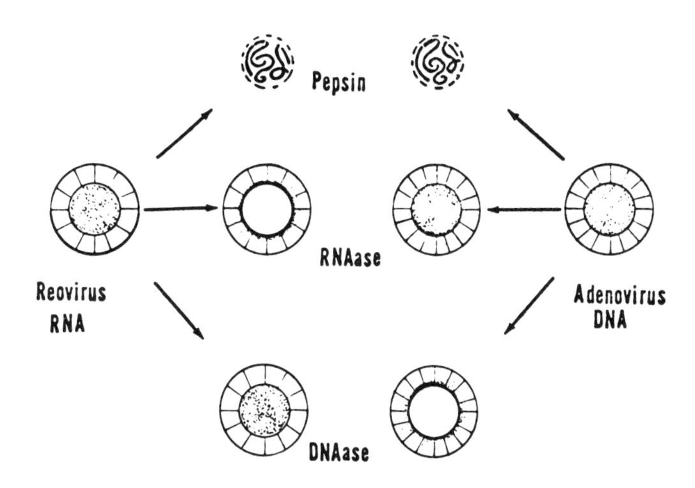

FIG. 8. Reaction of reoviruses and adenoviruses to differential treatments with pepsin to remove protein coat and with RNase, respectively, to remove nucleic acid core. From [13]: Chart II by permission.

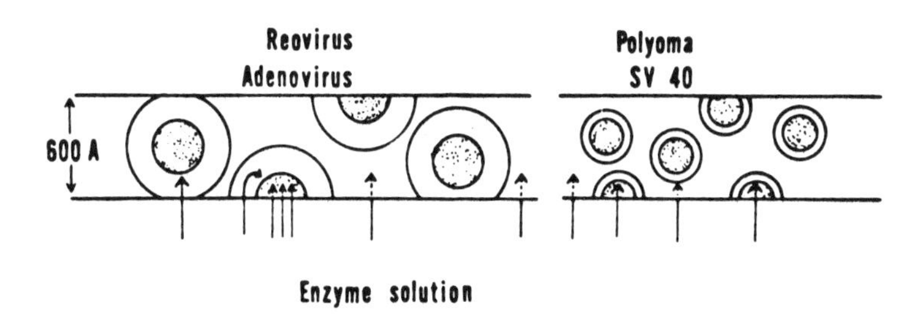

FIG. 9. The size of the virus is important in the effective penetration and reaction of enzyme solutions. From [13]: Chart I by permission.

Chapter 8

NEGATIVE STAINING FOR ELECTRON MICROSCOPY

M. L. DePamphilis*

Department of Biochemistry
University of Wisconsin
Madison, Wisconsin

*Present address: Department of Biological Chemistry, Harvard Medical School, Boston, Massachusetts.

I. INTRODUCTION

Negative staining refers to the fact that stain does not interact directly with the particle. A specimen is, instead, surrounded by an electron-dense stain (a soluble complex of a heavy metal). Since the paths of electrons traveling through a negatively stained specimen are relatively straight compared to the paths of electrons scattered in the stained area, the resulting specimen image appears as a light area (specimen) against a dark background (stain). This technique generally consists of allowing a sample to settle on the surface of a plastic film supported by a metal screen and then either first fixing the specimen in an aldehyde solution or directly applying a solution of a heavy metal salt. Only a thin film of stain solution is permitted to dry on the sample. The stain fails to penetrate protein molecules (i.e., positive staining) because the pH is such that charge repulsion occurs between stain and protein, and excess stain is immediately removed, thus limiting to 1 or 2 min the time that protein is soaking in stain solution. The result is crystals of stain filling cavities between protein aggregates and along the sides and bottom of the specimen where puddles of liquid formed on drying. The net effect is a topographical map of the specimen's surface. The extent of detail depends on the stain's intercalation.

For further discussion of the theory see papers by Horne [40,41], Finch and Klug [24], and Harris [30]. The advantages of negative staining over positive staining or metal shadowing are simplicity and rapidity, as well as a marked improvement in image detail.

II. METHODS

Excellent descriptions of materials and methods for sample preparation and electron microscopy are found in Kay [45], Horne [41], and Tikhonenko [69]. Although many variations in methods are found in the literature, the following procedure for negative staining is one that is applicable to most situations and gives excellent results.

A. Preparation of Formvar-Coated Grids

First, take precautions to avoid dust, smoke, grease, and oil vapors in the work area. Then clean 10 to 20 new 400-mesh copper grids (Ernest F. Fullam, Inc., P. O. Box 444, Schenectady, New York) for 20 min in distilled acetone. Let the grids dry on the surface of a hardened filter paper such as Whatman #50 in a covered petri dish. The copper grids or "screens" contain relatively large holes, which must be covered with a thin plastic film in order to support the specimen. Formvar (polyvinylformaldehyde) is presently the most generally used film and forms a stronger, but thicker, film than collodion or parlodion (both are nitrocellulose), which are also used.

To prepare a Formvar film, a glass microscope slide (3 x 1 in) is first cleaned with a nonchlorinated scouring powder such as Bon Ami (Standard Household Products, Corp., Holyoke, Massachusetts), and rinsed in deionized water. The use of detergents or caustic chemicals makes the Formvar very difficult to remove.

The slide is wiped dry with lens paper and lint is removed with compressed air from an EFFA Duster (Ernest F. Fullam, Inc.). The slide should be handled from one end only in order to keep fingers off the area to be used. The clean slide is immersed to 75% of its area in a 0.20% solution of Formvar (Ernest F. Fullam, Inc.) in distilled dichloroethane.* Excess Formvar is drained from the slide in a dichloroethane atmosphere which results in thinner, more uniform films than if the slide is ex-

---

*Store the Formvar solution in a wide-mouth glass-stoppered bottle placed in a desiccator containing silica gel. Add molecular sieves (Fisher Scientific Co., 4-Å pore size) to the Formvar solution to absorb water and other impurities. The stock solution is good for about 4 months. If the solvent is not anhydrous or the Formvar is allowed to dry in humid air, holes will form as the film dries. Once dry, the film resists water and most organic solvents.

posed immediately to air allowing the dichloroethane to rapidly evaporate. This is conveniently done in a 3 x 13 cm test tube preequilibrated with dichloroethane contained in a 5-ml beaker inside the tube. The slide is introduced vertically, so that the same edge that first entered the Formvar solution rests on top of the beaker. After 0.5 to 1 min, the slide is removed to dry in air for 3-5 min.

The Formvar film is then stripped from the slide and floated on the surface of deionized water as follows. Fill a glass dish (15.0 cm in diameter, 7.5 cm deep) to the brim with deionized water. Remove dust from the water surface by adding 5 drops of a 1% solution of collodion (Merck) in amyl acetate. The drops spread over the surface and form a film upon evaporation of the amyl acetate. Remove the collodion film, which now contains the surface dust, with a glass rod. Next, use a razor blade to score the Formvar film on one side of the glass slide 3 mm from each edge of the slide. Gently fog the film's surface with your breath (like fogging a window on a cold day) to loosen the film.* With the scored film surface up, slowly dip the slide

---

*People who are heavy smokers may find that passage of the film through the cooler vapors from boiling water is more satisfactory because black carbon specks from their breath can contaminate the Formvar.

into the dish of clean water at a 30-deg angle. The film will float free. By looking at light reflected from the film's surface, you can find a thin area that is smooth, homogeneous, and nearly transparent.

Using a fine-tipped forceps, pick up the grids by their edge and then drop them in an orderly array onto the thin part of the film so that their dull sides face the Formvar. The film sticks more firmly to the dull side of the grid and is thus more stable in the electron microscope.

A simple method for retrieving the Formvar-coated grids is illustrated in Fig. 1. An alternative method to cover the grids with a plastic film is described by Bradley [11, p. 62]. (Note that a sintered-glass funnel mounted on a side-armed Erlenmeyer flask can be used in place of the modified petri dish [11].)

Resolution achievable in electron microscopy will be affected by the thickness of the supporting film. To overcome this problem, films containing many small holes can be easily prepared [11,29, 42]. The stain will form a thin film over these holes, entrapping samples of the specimen and resulting in greater contrast and thus higher resolution. However, one needs a high concentration of specimen to ensure finding good examples over the holes. Finch and Klug [25] have compared images of turnip-yellow mosaic virus seen over "holes" with those seen resting on the film. Other examples are F-actin [32], F-actin complexed with myosin-head subunits [59], and bacterial flagellar filaments [52].

B. Application of a Carbon Film

A carbon layer is usually evaporated over the Formvar film to give it additional stability and prevent its tearing and drifting under the electron beam. This is particularly true when higher magnifications and, therefore, higher-beam currents are used. Unfortunately, the carbon layer is hydrophobic and repels ionic stains, which results in uneven spreading of the stain and the formation of stain puddles. This problem is not serious when crude cell lysates are examined since an excess of soluble protein coats the carbon surface and helps spread the stain. When purified materials are studied, either the concentration of specimen material may be raised or 0.01% bovine serum albumin may be added to silicotungstate or phosphotungstate stains [26,67]. However, albumin precipitates in uranyl acetate, pH 4.5.

The hydrophobic effects of the carbon layer are circumvented if carbon is instead evaporated on the back of the Formvar film. Each Formvar-coated grid is inverted so that the Formvar side is facing the glass. A thin carbon film is applied by graphite evaporation [11,41]. The layer of carbon should be light enough so that when a grid is removed, only a faint outline of the grid is seen on the glass. The grids now have a hydrophilic Formvar film on their dull side and a carbon film on their shiny side. The grids can be stored in a covered petri dish.

C. Negative Staining

Hydrocarbons that are absorbed in the vacuum evaporator or on standing in air will also contribute a hydrophobic character to the Formvar surface. One removes them just prior to use by gripping the edge of the grid in a locked forceps and touching the Formvar-coated side to the surface of petroleum ether for 5 sec, then drawing off the excess by touching the edge of the grid with a piece of filter paper. A drop of the sample is placed on the clean Formvar surface and left until a sufficient amount of specimen has settled, as determined by trial and error. For a 0.5-mg/ml solution of bacterial flagella, this requires about 1 min. Once the sample is applied, the Formvar surface should not be allowed to dry until the end of the procedure, otherwise poor staining or specimen distortion may result. The Formvar surface is then touched to the surface of 20% Formalin or glutaraldehyde for 30-60 sec and then to deionized water for 15 sec. This process serves to preserve specimen structure ("fixation") and to wash out salts that may be present in the sample. Excess water is removed with filter paper and a drop of stain solution is placed on the grid.* Excess stain is re-

---

*Commonly used stains are solutions of 1% uranyl acetate in deionized water, pH 4.5, or 1% potassium phosphotungstate titrated to pH 6.8-7.2 with KOH. The solutions are stored at 5°C and an aliquot is filtered through Watman #50 filter paper just before use to remove insoluble materials that may accumulate. For premoved within 5 sec with a piece of filter paper. Since the quality of negative staining depends on the amount of stain that dries on the grid, one must empirically adjust the quantity of stain either by staining a series of identical samples and removing different amounts of stain, or by using different concentrations of stain. When stain penetration is desired, the stain is allowed to remain up to 1 min before removal of excess solution. The grids are then air dried, although some workers [7,10,68] prefer to let them dry under vacuum in the electron microscope to ensure complete drying and to reduce possible damage from the high surface tension of water. Detailed studies of virus structure generally detect some distortion via particle flattening presumably during drying [24,25].

However, the best method for minimizing distortion that results from drying is the critical-point drying technique [6]. Although it is designed for drying samples destined for metal shadowing and, therefore, unstained, the principle is applicable to negatively stained materials. Absolute ethanol is substituted for the water on the grid, followed by amyl acetate in exchange for the alcohol, and finally liquid carbon dioxide replaces the amyl acetate. The temperature is raised above 31.1°C where the surface tension of carbon dioxide vanishes and CO_2 is removed as a gas. The sample is thus dried at a reasonable temperature without passing through a liquid-air interface and the surface tension forces that go with it. If equipment for working with

---

paration of other stains, see the references in Table 1, Abrams and Davis [1], and Horne [41].

liquid CO_2 is not available, one may still find improvement in specimen structure after drying in amyl acetate.

Stained specimens should be examined immediately, since image quality will deteriorate with time, owing to absorption of particles and oil from the air. In addition, Harris and Westwood [31] found that the appearance of vaccinia virus changed markedly in a few hours as a result of continued penetration of stain.

Two other methods to apply a sample to the grid can be useful. One method is to put a drop of sample on a collodion membrane spread over semidry agar and to let the water dialyze into the agar, which leaves the sample on the membrane [46,70]. Another method is to mix the sample with a stain and apply it as a fine spray, as described in detail by Horne [39]. Although qualitative differences in the settling of particles on a grid are generally not a problem [70], the value of these techniques is that they ensure collecting a representative sample, increase specimen concentration on the grid, and permit counting the number of particles in the original solution (e.g., virus particle titer) [39].

III. DISCUSSION

The following material is intended to give the novice a feeling for negative staining by citing examples of its application and noting pitfalls that can occur. The generalizations are only guidelines and, in the end, practice will prove the best teacher.

A. Choice of Stain

Table 1 provides examples of materials that have been studied by negative staining. They illustrate the variation in appearance of specimens and the amount of detail one can expect from this technique by the use of various stains.

TABLE 1

Examples of Negatively Stained Materials

Materials	Stains[a]	References
Bacterial cells	UA, NaST	9, 10, 67
Bacterial cell walls	UA, KPT, AM	35, 60
Bacterial flagellar filaments	UA, KPT	52
Bacterial flagellar basal ends and attachment to cell envelope	UA, KPT	2-4, 16-18, 34
Protozoan flagellar microtubules	UA, KPT	27, 28, 36
Structures attached to microtubules	UA	37
Actin, myosin, actin-meromyosin	UA, UF	32, 43, 59
Pyocin (a rod-shaped bacteriocin)	UA	33
Antigen-antibody complexes	UA	8, 74

TABLE 1 cont'd.

Materials	Stains[a]	References
Ribosomes	UA, KPT, NaT	13, 42
Large enzyme complexes	KPT	63
Small cylindrical proteins	UA, KPT	30
Small enzymes	UA, KPT, AM	38, 48, 49, 72
Bacteriophage and components	UA, KPT, UF, NaST	47, 58, 67-69
Small spherical viruses	UA, KPT	24, 25
Adenovirus	NaST	71
Rod-shaped viruses	UA, KPT, UF	19, 22, 51, 62, 73
Mitochondrial inner membrane	KPT	21
Lipid micelles	KPT, AM, NaT, CaPT, LiT	53, 65

[a]UA = uranyl acetate, UF = uranyl formate, KPT = potassium phosphotungstate, CaPT = calcium phosphotungstate, NaST = sodium silicotungstate, AM = ammonium molybdate, NaT = sodium tungstate, LiT = lithium tungstate.

With isolated proteins, uranyl acetate (UA) generally gives greater penetration, contrast, and detail than does phosphotungstate. In addition, the attachment of flagella or bacteriophage to untreated bacterial cell surfaces can be studied as well as the contour of these surfaces [9,10,67]. However, UA tends to accumulate in empty vesicles such as virus capsids and isolated membrane fragments, resulting in a potential loss of detail.

Potassium phosphotungstate (KPT) is most useful where surface detail is desired (e.g., spherical viruses) and in the study of unpurified materials in cell lysates. The KPT distributes more evenly over the specimen and thus is less likely to mask the presence of small structures in close proximity to larger ones.

Uranyl formate (UF) is smaller and forms denser crystals than do other stains and, in theory, should give greater penetration and contrast [51]. The UF has been found superior in studies of rod-shaped viruses [22].

Ammonium molybdate (AM) and sodium silicotungstate (NaST) are similar to KPT in their general staining characteristics. However, unlike KPT, the AM does not inhibit a galactosyl transferase reaction whose activity depends upon binding of the enzyme to lipopolysaccharide-phospholipid membrane complexes [65], which suggests that AM is less disruptive of macromolecular structures than is KPT. The NaST stain is widely used on bacteriophage and appears to be better than KPT or UA for the definition of tail fibers and pins [67,68].

The appearance of materials can be greatly modified by the choice of stain, and therefore, various stains should be tested both with and without prior fixation of the material. Huxley [43] reported that KPT destroyed the structure of actinmeromyosin complexes while UA gave excellent detail, and Hopkins [37] noted that tiny projections attached to microtubules were seen with UA but were damaged and partially missing with KPT. Van Bruggen et al. [12] reported that KPT caused disruption of the structure of hemocyanin, whereas UA did not. On the other hand, Grimstone and Klug [28] found KPT superior to UA for observing subunit arrangements in microtubules, and Abram and Koffler [2] reported that UA caused partial or complete destruction of flagellar filaments, although this was probably owing to the acidity of the stain (pH 4.5) rather than to the stain itself. Tikhonenko [69] found more particles of phage 1 of *Bacillus mycoides* with intact sheaths when preparations were obtained with KPT than when UA was used. An extensive study by Abram and Davis [1] showed that different negative stains developed different aspects of the morphology of *Bdellovibrio bacteriovorus*. Variation of a stain's pH and ionic strength also influenced their results.

These effects might be minimized by prior fixation of the proteins with Formalin (37% formaldehyde with 10-15% methanol to prevent polymerization) or glutaraldehyde, which preserves the structure of proteins by crosslinking primary amino groups [57,64] while giving little increase in electron density. This treatment is apparently mild since many enzymes retain activity after aldehyde fixation, particularly when glutaraldehyde or hydroxyadipaldehyde is used [20,66]. Aldehyde fixation has been reported to be necessary in studies of ribosomes [13,42], bacteriophage [47], flagella [17], and enzymes [72].

The most extensive structural changes that result from the type of stain used, as well as from its pH and ionic strength, are seen in studies of lipid micelles that contain combinations of lecthin, cholesterol, and saponin [53], and less dramatically with phosphatidyl ethanolamine and lipopolysaccharide combinations [65]. Aldehyde fixation would be unlikely to reduce these changes.

B. Interpretation of Electron Micrographs

Finch and Holmes [22] give an excellent discussion of this topic as it applies to virus structure. The following are notable comments.

Specimen appearance is altered by the thickness of the surrounding stain. Harris [30] has demonstrated how stain thickness changes the appearance of a small cylindrical protein released from erythocyte ghosts. Crawford et al. [15] found that polyoma capsids could appear full or empty, depending on the stain concentration. I have observed fields of "intact" flagella with only the hook-basal-body complexes visible because the thinner filaments were covered with stain. Moody [58] has analyzed the structure of T4-phage tail sheaths by studying specimens resting in varying amounts of stain and at different angles to the grid.

Stain penetration also determines the apparent size of a specimen [26]. For example, the distance between centers of adjacent subunits in flagella or microtubules is often greater than the apparent diameter of the subunit. Diameters of particles such as flagella, virus, and ribosomes should be confirmed by positive staining or metal shadowing.

The presence of salts or sugars (e.g., sucrose) in a sample may result in the formation of crystals on the grid that can be mistaken for structures. One should, therefore, minimize the concentration of salts and sugars present, as well as always wash the specimen in distilled water after it has been mounted on the grid but before it is stained. This is ordinarily done during the aldehyde fixation step, but one may sometimes wish to study unfixed materials. Alternatively, one can examine a duplicate of the sample solution minus the specimen, or change the concentration of the salt or sugar and note any changes in specimen appearance.

Finally owing to great depth of field, electron microscopic images of negatively stained materials such as viruses and flagella are ordinarily composed of superimposed images of the top and bottom surfaces of the specimen [22,50,52]. The result is often a complex hazy pattern that appears disordered as a result of out-of-register contributions from the two images. In addition, the contributions from each side can vary from predominantly one-sided images (1-5% [24]) to images with equal contributions from both sides. Finch and Klug [24,25] discuss the interpretation of virus structure from two-sided composite images and demonstrate the value of photographing the specimen from different angles to give a stereoscopic image. They conclude that the side of the specimen facing the grid gives the dominant image.

One can also try to identify images with plaster of paris (the negative stain), prepare an x-ray image from them to give a "negatively stained" image of the model, and then compare these images with electron micrographs of the specimen [14].

C. Image-Enhancement Techniques

The amount of information obtained from an electron micrograph could, in principle, be increased if the noise level (factors contributing to background images) were reduced. Klug and DeRosier [50] and Finch and Holmes [22] discuss a sophisticated technique for reconstructing one-sided images by first obtaining what amounts to a diffraction pattern from an electron micrograph and then optically filtering out all signals except those that appear to arise from one side of the specimen before reconstructing the image by reversing the process. An excellent description of the instrument used to obtain optical diffraction patterns and of their interpretation is given by Markham [54].

With specimens having rotational symmetry, enhancement is achieved simply by placement of a photographic negative of the specimen in an enlarger and focusing it onto photographic paper mounted on a turntable [55]. The exposure time t required for a high-contrast image is determined. A fresh paper is then rotated 360°/n and exposed for t/n sec, where n is an integral number. This is continued for n exposures. If n is a fundamental periodicity in the specimen, this periodicity will be reinforced. The same principles can be applied also to linear objects [56]. However, these techniques must be used with caution since artifacts are easily reinforced [5]. Agrawal et al. [5] suggest that the technique be restricted to the enhancement of structures that are already visible, or at least have known symmetry and a visible center of symmetry in the unrotated micrograph. For example, Finch et al. [23] examine the rotational symmetry of tobacco mosaic virus both by x-ray diffraction and by superposition of six different electron-microscope images of viral disks. They find 17-fold symmetry by x-ray rather than the 16-fold symmetry exposed with the rotational enhancement technique.

A third method of image enhancement is to superimpose many different images of a specimen. The first of n images is printed and this print is used to center the remaining images as follows. A negative is placed in the enlarger and the platform that holds the print in place is moved until the new image is superimposed on the printed one. The print is replaced by an undeveloped sheet of printing paper and exposed for approximately 1/n the time required for a high-contrast image. The process is repeated until about six different images are exposed on the same print. Examples of the use of this technique can be seen with tobacco mosaic virus disks [23] and glutamine synthetase [72].

Finally, contrast can be increased by reversal of the image once or twice. A useful high-contrast film for this purpose is

DuPont Ortho A Litho sheet film (#COA7). Standard electron microscope sheet films such as Kodlith LR (#2572) or Kodak Electron Microscope film (#4489) also can be used. The procedure is to place the emulsion side of the film against the emulsion side of the negative, clamp them firmly between two sheets of glass, and expose them over a light box. This technique works best with shadowed materials such as DNA, but it may prove useful with certain negatively stained specimens. The results of this method can be seen with flagellar filaments [52], F actin [32], and catalase crystals [49]. Since this method enhances only images that already show the most contrast, artifacts can be easily emphasized.

ACKNOWLEDGMENTS

I thank Dr. Donna Kubai, Dr. Robert Mesibov, and Dr. Gary Borisy for reading this manuscript and offering many helpful suggestions.

REFERENCES

1. D. Abram and B. K. Davis, J. Bacteriol., 104, 948 (1970).

2. D. Abram and H. Koffler, J. Mol. Biol., 9, 168 (1964).

3. D. Abram, H. Koffler, and A. E. Vatter, J. Bacteriol., 90, 1337 (1965).

4. D. Abram, J. R. Mitchen, H. Koffler, and A. E. Vatter, J. Bacteriol., 101, 250 (1970).

5. H. O. Agrawal, J. W. Kent, and D. M. MacKay, Science, 148, 638 (1965).

6. T. F. Anderson, in Physical Techniques in Biological Research (G. Oster and A. W. Pollister, eds.), Vol. III, Academic Press, New York, 1956, p. 177.

7. T. F. Anderson and R. Stephens, Virology, 23, 113 (1964).

8. S. Askura, G. Eguchi, and T. Iino, J. Mol. Biol., 35, 227 (1968).

9. M. E. Bayer, J. Gen. Microbiol., 46, 237 (1967).

10. M. E. Bayer and T. F. Anderson, Proc. Natl. Acad. Sci. (U. S.), 54, 1592 (1965).

11. D. E. Bradley, in Techniques for Electron Microscopy (D. H. Kay, ed.), Blackwell Scientific Publications, Oxford, 1965, p. 58.

12. E. F. J. van Bruggen, E. H. Weibenga, and M. Gruber, Biochim. Biophys. Acta, 42, 171 (1960).

13. V. I. Bruskov and N. A. Kiselev, J. Mol. Biol., 37, 367 (1968).

14. D. L. D. Caspar, J. Mol. Biol., 15, 365 (1966).

15. L. V. Crawford, E. M. Crawford, and D. H. Watson, Virology, 18, 170 (1962).

16. M. L. DePamphilis, J. Bacteriol., 105, 1184 (1971).

17. M. L. DePamphilis and J. Adler, J. Bacteriol., 105, 384 (1971).

18. M. L. DePamphilis and J. Adler, J. Bacteriol., 105, 396 (1971).

19. A. C. H. Durham, J. T. Finch, and A. Klug, Nature, 229, 37 (1971).

20. J. L. E. Ericsson and P. Biberfeld, Lab. Invest., 17, 281 (1968).

21. H. Fernandez-Moran, T. Oda, P. V. Blair, and D. E. Green, J. Cell. Biol., 22, 63 (1969).

22. J. T. Finch and K. C. Holmes, in Methods in Virology (K. Maramorosch and H. Koprowski, eds.), Vol. III, Academic Press, New York, 1967, p. 351.

23. J. T. Finch, R. Leberman, C. Yu-Shang, and A. Klug, Nature, 212, 349 (1966).

24. J. T. Finch and A. Klug, J. Mol. Biol., 13, 1 (1965).

25. J. T. Finch and A. Klug, J. Mol. Biol., 15, 344 (1966).

26. A. M. Glauert, in Quantitative Electron Microscopy (G. F. Bahr and E. Zeitler, eds.), Williams and Wilkins, Baltimore, Maryland, 1965, p. 331.

27. A. V. Grimstone, in Formation and Fate of Cell Organelles (K. B. Warren, ed.), Academic Press, New York, 1967, p. 219.

28. A. V. Grimstone and A. Klug, J. Cell. Sci., 1, 351 (1966).

29. W. J. Harris, Nature, 196, 499 (1962).

30. J. R. Harris, J. Mol. Biol., 46, 329 (1969).

31. W. J. Harris and J. C. N. Westwood, J. Gen. Microbiol., 34, 491 (1964).

32. J. Hanson and J. Lowy, J. Mol. Biol., 6, 46 (1963).

33. T. B. Higerd, C. A. Baechler, and R. S. Berk, J. Bacteriol., 98, 1378 (1969).

34. J. F. M. Hoeniger, W. Van Iterson, and E. N. Van Zanten, J. Cell. Biol., 31, 603 (1966).

35. S. C. Holt and E. R. Leadbetter, Bacteriol. Rev., 33, 346 (1969).

36. D. E. Hooks, J. Randall, and J. M. Hopkins, in Formation and Fate of Cell Organelles (K. B. Warren, ed.), Academic Press, New York, 1967, p. 115.

37. J. M. Hopkins, J. Cell Sci., 7, 823 (1970).

38. R. W. Horne, in Quantitative Electron Microscopy (G. F. Bahr and E. H. Zeitler, eds.), Williams and Wilkins, Baltimore, Maryland, 1965, p. 316.

39. R. W. Horne, in Techniques for Electron Microscopy (D. H. Kay, ed.), Blackwell Scientific Publications, Oxford, 1965, p. 311.

40. R. W. Horne, in Techniques for Electron Microscopy (D. H. Kay, ed.), Blackwell Scientific Publications, Oxford, 1965, p. 328.

41. R. W. Horne, in Methods in Virology (K. Maramorosch and H. Koprowski, eds.), Vol. III, Academic Press, New York, 1967, p. 521.

42. H. E. Huxley and G. Zubay, J. Mol. Biol., 2, 10 (1960).

43. H. E. Huxley, J. Mol. Biol., 7, 281 (1963).

44. I. Iino, Bacteriol. Rev., 33, 454 (1969).

45. D. H. Kay, Techniques for Electron Microscopy, Blackwell Scientific Foundation, Oxford, 1965.

46. E. Kellenberger and W. Arber, Virology, 3, 245 (1957).

47. E. Kellenberger, A. Bolle, E. Boy de la Tour, R. H. Epstein, N. C. Franklin, N. K. Jerne, A. Reale-Scafati, J. Se'chaud, I. Bendet, D. Goldstein, and M. A. Lauffer, Virology, 26, 419 (1965).

48. N. A. Kiselev, F. Y. Lerner, and N. B. Livanova, J. Mol. Biol., 62, 537 (1971).

49. N. A. Kiselev, C. L. Shpitzberg, and B. K. Vainshtein, J. Mol. Biol., 25, 433 (1967).

50. A. Klug and D. J. DeRosier, Nature, 212, 29 (1966).

51. R. Lederman, J. Mol. Biol., 13, 606 (1965).

52. J. Lowy and J. Hanson, J. Mol. Biol., 11, 293 (1965).

53. J. A. Lucy and A. Glauert, in Formation and Fate of Cell Organelles (K. B. Warren, ed.), Academic Press, New York, 1967, p. 251.

54. R. Markham, in Methods in Virology (K. Maramorosch and H. Koprowski, eds.), Vol. 4, Academic Press, New York, 1968, p. 504.

55. R. Markham, S. Frey, and G. J. Hills, Virology, 20, 88 (1963).

56. R. Markham, J. H. Hitchburn, G. J. Hills, and S. Frey, Virology, 22, 342 (1964).

57. G. E. Means and R. E. Feeney, Biochemistry, 7, 2192 (1968).

58. M. F. Moody, J. Mol. Biol., 25, 167 (1967).

59. P. B. Moore, H. E. Huxley, and D. J. DeRosier, J. Mol. Biol., 50, 279 (1970).

60. R. G. E. Murray, Can. J. Microbiol., 9, 381 (1963).

61. J. Naginton, A. A. Newton, and R. W. Horne, Virology, 23, 461 (1964).

62. R. E. Offord, J. Mol. Biol., 17, 370 (1966).

63. L. J. Reed and D. J. Cox, Ann. Rev. Biochem., 35, 57 (1966).

64. F. M. Richards and J. R. Knowles, J. Mol. Biol., 37, 231 (1968).

65. L. Rothfield and R. W. Horne, J. Bacteriol., 93, 1705 (1967).

66. D. D. Sabatini, K. G. Bensch, and R. J. Barrnett, J. Cell. Biol., 17, 19 (1963).

67. S. Z. Schade, J. Adler, H. Ris, J. Virology, 1, 599 (1967).

68. L. D. Simon and T. F. Anderson, Virology, 32, 279 (1967).

69. A. S. Tikhonenko, Ultrastructure of Bacterial Viruses, Plenum Press, New York, 1970.

70. C. M. To, E. Kellenberger, and A. Eisenstack, J. Mol. Biol., 46, 493 (1969).

71. R. C. Valentine and H. G. Pereiro, J. Mol. Biol., 13, 13 (1965).

72. R. C. Valentine, B. M. Shapiro, and E. R. Stadtman, Biochemistry, 7, 2143 (1968).

73. R. C. Williams and H. W. Fisher, J. Mol. Biol., 52, 121 (1970).

74. M. Yanagida and C. Ahmad-Zadeh, J. Mol. Biol., 51, 411 (1970).

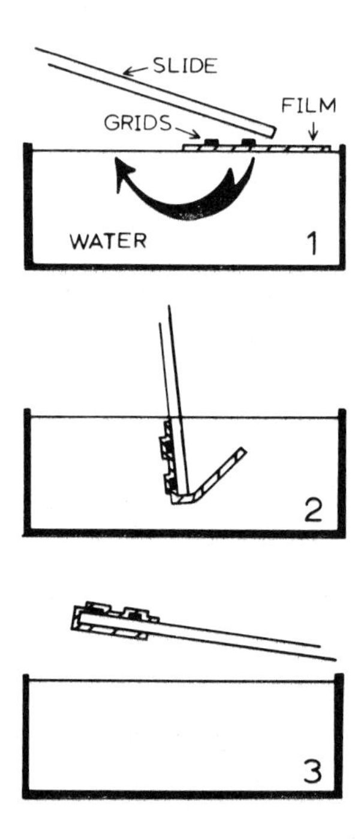

FIG. 1. The Formvar film is retrieved with grids attached by smooth passage of a clean microscope slide over the film as illustrated (steps 1, 2, and 3). Upon contact the film sticks tightly to the slide. As many as 30 grids can be conveniently prepared this way.

AUTHOR INDEX

Numbers in brackets are reference numbers and indicate that an author's work is referred to although his name is not cited in the text. Underlined numbers give the page on which the complete reference is listed.

A

B

SUBJECT INDEX